CLIMATE CHANGE AND ITS IMPACT ON THE HORTICULTURIST COMMUNITY

About the Authors

Dr. Jagriti Mehta (b. 1991), (B.Sc. Gold Medalist, M.Sc. Gold Medalist, Ph.D.) is an anthropologist with keen interest in the field of socio-cultural, climate and ecological anthropology. Her other research interest includes gender studies, medical anthropology and applied anthropology. She has done her Ph.D. from Panjab University, Chandigarh as a UGC-JRF and UGC-SRF. Currently she is a Guest Faculty in Department of Anthropology, Panjab University, Chandigarh. She has published 7 research articles and have attended 14 conferences.

Dr. Anil Kishore Sinha (b. 1958), did his M.Sc. and Ph.D. from Ranchi University, Ranchi. Currently he is a Professor in Department of Anthropology, Panjab University, Chandigarh. He has been teaching Social Anthropology since 1987. He has more than 140 research publications in national and international journals. He has written 15 books. He has completed 7 major research projects funded by different government agencies. His research interests include Social Anthropology, Tribal Studies, Gender Studies, Anthropology of Food, Medical Anthropology and Social Impact Assessment.

Dr. Maninder Kaur (b. 1972), did her M.Sc (Hons.), Ph.D. in Physical Anthropology from Panjab University, Chandigarh. Her area of specialization is biological Anthropology (Human Growth and Nutrition, Gerontology). She was awarded JRF by University Grants Commission in year 1999. She has successfully completed four major research projects funded by Department of Science and Technology, New Delhi. She is presently working on a major research project sponsored by DST, New Delhi. She is a member of Program Advisory Committee for National Science and Technology Management Information System Program of Department of Science and Technology, New Delhi. She has published many research papers in national and international journals of repute. She has also presented research papers at various conferences in India and abroad. Currently, she is chairing the Department of Anthropology and deputy coordinator of UGC CAS-II programme.

CLIMATE CHANGE AND ITS IMPACT ON THE HORTICULTURIST COMMUNITY

An Ethnographic Study

Jagriti Mehta
Anil Kishore Sinha
Maninder Kaur

CONCEPT PUBLISHING COMPANY PVT. LTD.
NEW DELHI - 110 059

ISBN: 978-93-5594-661-4

First Published 2024

Published and Printed by

Concept Since 1974

Concept Publishing Company Pvt. Ltd.
A/15-16, Commercial Block, Mohan Garden
New Delhi-110059 (India)
T: +91-11-41101460
E: publishing@conceptpub.com, *W*: www.conceptpub.com

Cataloging in Publication--Courtesy: D.K. Agencies (P) Ltd. <docinfo@dkagencies.com>

Mehta, Jagriti, 1991- **author.**
Climate change and its impact on the horticulturist community : an ethnographic study / Jagriti Mehta, Anil Kishore Sinha, Maninder Kaur.
pages cm
Study conducted in Kotkhai Tehsil of District Shimla, Himachal Pradesh, India.
Includes bibliographical references and index.
ISBN 9789355946614

1. Horticultural crops--Climatic factors--India--Kotkhai (Tehsil) 2. Human beings--Effect of climate on--India--Kotkhai (Tehsil) 3. Human ecology--India--Kotkhai (Tehsil) 4. Horticulturists --India--Kotkhai (Tehsil)--Economic conditions. 5. Horticulturists--India--Kotkhai (Tehsil)--Social conditions. 6. Horticulturists--India--Kotkhai (Tehsil)--Attitudes. I. Sinha, Anil Kishore, author. II. Maninder Kaur, 1972- author. III. Title.

LCC S600.64.I4M44 2024 | DDC 630.2515 23

Preface

As human beings, we have a unique relationship with the environment around us. Our actions, beliefs, and social structures shape and are shaped by the natural world in which we live. Anthropology, as a discipline, seeks to understand this intricate relationship between humans and their environment by studying cultures, societies, and communities in their ecological contexts.

Climate change, with its global impact on ecosystems, weather patterns, and natural resources, has emerged as one of the most pressing issues of our time. The impact of climate change on human communities, especially those dependent on the environment for their livelihoods, is a crucial area of inquiry.

In this book, we will explore the impact of climate change on the horticultural community of Himachal Pradesh, a state located in the northern part of India. Himachal Pradesh is known for its scenic beauty and rich cultural heritage, but it is also home to a large number of farmers who rely on horticulture for their livelihood. By exploring the socio-cultural and ecological dimensions of climate change, the book provides insights into the impact of climate change on various livelihood assets and complex ways in which human societies are responding to and coping with environmental challenges.

In recent years, the state has witnessed changes in climate patterns that have had a significant impact on the horticultural community. The farmers are facing challenges such as irregular rainfall, declining and erratic snowfall, rising temperatures, and increased incidence of pests and diseases, all of which are affecting crop yields and quality. Through ethnographic case studies and interviews with the horticultural community members, the book highlights the lived experiences of those on the frontlines of climate change in Himachal Pradesh. It shows how climate change is affecting the cultural practices, social relationships, and economic activities of the horticultural community in the region.

The book also examines the role of institutions, governance structures, and policies in shaping the response of the horticultural community to climate change. It underscores the importance of community participation, local knowledge, and traditional practices in devising effective adaptation and mitigation strategies.

This book is intended to be a valuable resource for students, research scholars, academicians, social scientists, environmentalists, botanists, horticulturists, NGOs, policymakers, and anyone interested in understanding the anthropological dimensions of climate change and its impact on human communities. It is anticipated that the insights and perspectives presented in this book will contribute to the ongoing dialogue and efforts towards creating a sustainable and just future for all.

Jagriti Mehta
Anil Kishore Sinha
Maninder Kaur

Acknowledgements

We would like to express our sincere gratitude to everyone who has supported us throughout the writing of this book. Firstly, we would like to thank our families for their unwavering support and encouragement.

We would like to extend our sincere thanks to the respondents of the research on which this book is based, without them the research work would not have been possible. We are grateful for the data provided by the respondents, people of villages Bhawana, Kiari, Kufar Bag and Jashla.

We extend our gratitude towards the University Grants Commission (UGC), New Delhi and Panjab University, Chandigarh. Their financial support, research resources, and academic forums have provided us with the necessary foundation to pursue our scholarly goal.

We would also like to express our thanks to our editors, who have been instrumental in shaping this book into its final form. Their insightful comments and suggestions have been invaluable, and we are grateful for their hard work and dedication.

We would like to pay our humble gratitude to the Higher Power for blessing us with health and good fortune throughout our journey.

Last but not the least, we are forever grateful to the discipline of Anthropology for making us better people.

Jagriti Mehta
Anil Kishore Sinha
Maninder Kaur

Contents

Preface v
Acknowledgements vii
List of Tables, Figures and Plates xv
List of Abbreviations xix
Glossary xxi

1. Introduction **1**
1.1 Climate change 1
1.2 Anthropology of climate change 3
1.2.1 History of anthropology of climate change 4
1.2.2 Importance of anthropological approach in studying climate change 5
1.3 Livelihood 5
1.4 The nexus between climate change and livelihood 8
1.5 The climate and livelihood of the area of study 9
1.6 Statement of the problem 11
1.7 Rationale of the problem 12
1.8 Review of literature 13
1.8.1 Anthropology and climate change 14
1.8.2 Climate change trends 15
1.8.3 Impact of climate change on livelihood 18
1.8.4 Impact of climate change on agricultural livelihood 24
1.8.5 Adaptation made to combat and mitigate climate change 25
1.9 Aim and objectives 27
1.10 Research methodology 27
1.10.1 Universe and unit of study 28

1.10.2 Rationale of the selection of study area 28
1.10.3 Rapport establishment 29
1.10.4 Sampling 30
1.10.5 Details of field visits 32
1.10.6 Ethical considerations 32
1.10.7 Data collection 33
1.10.8 Data analysis 37
1.10.9 Problems encountered 42
1.11 Significance of the study 43
1.12 Limitations of the study 44
1.13 Chapterization 44

2. Area and People 46
2.1 India 46
2.2 Himachal Pradesh 47
2.2.1 Topography 48
2.2.2 Administrative units 49
2.2.3 Demography 50
2.2.4 Economy 50
2.2.5 Climate 51
2.3 Shimla 51
2.3.1 Topography 52
2.3.2 Administrative units 52
2.3.3 Demography 52
2.3.4 Economy 53
2.3.5 Climate 53
2.4 Kotkhai tehsil 54
2.4.1 History 55
2.4.2 Demography 55
2.4.3 Economy 56
2.4.4 Religion 57
2.4.5 Rituals, customs and festivals 58
2.4.6 Language 59
2.4.7 Dress and ornament 59
2.4.8 Family structure 59
2.5 Field area 59
2.5.1 Bhawana 60
2.5.2 Kiari 61

2.5.3 Kufar Bag 62
2.5.4 Jashla 63

3. Climate Change and People's Perception 65
3.1 Analysis of climatic data of the study area 65
3.1.1 Temperature 65
3.1.2 Rainfall 68
3.1.3 Snowfall 74
3.2 People's perception about climate change and its impact on their day to day lives 76
3.2.1 Temperature 76
3.2.2 Rainfall 77
3.2.3 Snowfall 79
3.2.4 People's perception about cause of climate change 81
3.2.5 Factors affecting the perception about climate change 85
3.2.6 Impact of climate change on day to day lives 88
3.3 Comparison of people's perception and climate data 91

4. Impact of Climate Change on Crop Varieties, Calendar and Crop Production 93
4.1 Crop documentation 94
4.1.1 Evolution of the present horticulture practices 94
4.1.2 Changes in crop varieties with changing climate 95
4.2 Crop calendar 97
4.3 Crop production 105
4.3.1 People's perception 105
4.3.2 Crop production based on secondary data 109

5. Impact of Climate Change on the Livelihood: An Emic View 125
5.1 Impact of climate change on natural assets of livelihood 127
5.1.1 Soil/Land 127
5.1.2 Water 129
5.1.3 Biodiversity 136

5.2 Impact of climate change on human assets of livelihood 141
5.2.1 Household members 142
5.2.2 Active labour 143
5.2.3 Education 145
5.2.4 Health 145
5.2.5 Knowledge and skills 151
5.3 Impact of climate change on the social assets of livelihood 152
5.4 Impact of climate change on physical assets of livelihood 154
5.5 Impact of climate change on financial assets of livelihood 158
5.6 Impact of climate change on horticulture 159
5.6.1 Temperature 162
5.6.2 Rainfall 165
5.6.3 Snowfall 167
5.6.4 Extreme weather events 169
5.6.5 Increase in use of chemicals 172

6. Adaptation Strategies and Government's Efforts to Combat Climate Change 175
6.1 Adaptation strategies acquired by the people to overcome the impact of climate change 175
6.1.1 Adaptation strategies acquired by the people for the lower altitude village Bhawana 176
6.1.2 Adaptation strategies acquired by the people for the lower-middle altitude village Kiari 180
6.1.3 Adaptation strategies acquired by the people for the upper-middle altitude village Kufar Bag 182
6.1.4 Adaptation strategies acquired by the people for the high-altitude village Jashla 186
6.1.5 Changing dynamics of social relations as an adaptation to combat climate change 188
6.1.6 Steps to mitigate the impact of climate change 190

6.2 The government efforts being taken to combat and mitigate climate change 192

6.2.1 Schemes and projects being implemented in the field area as per the respondents 194

6.2.2 Issues related to the schemes faced by the respondents 196

7. Summary and Conclusion **199**

7.1 Summary 199

7.2 Conclusion 216

Appendix 219

References 222

Index 237

List of Tables, Figures and Plates

Tables

1.1 Types of livelihood assets 6
1.2 Livelihood contexts 7
1.3 Criteria for sample size selection 31
2.1 Statistical detail of Himachal Pradesh 50
2.2 Statistical detail of Tehsil Kotkhai (Census, 2011) 56
2.3 Religion-wise Population of Tehsil Kotkhai 57
2.4 Demographic data of village Bhawana 61
2.5 Demographic data of village Kiari 62
2.6 Demographic data of village Kufar Bag 63
2.7 Demographic data of village Jashla 63
3.1 The traditional calendar with expected and actual weather events 87
4.1 List of crops grown for commercial purpose 94
4.2 Traditional and new varieties of crop documented in the field area 97
4.3 Changes in crop calendar (apple and pear) from green tip stage to harvesting time observed in the selected villages 101
4.4 Changes in crop calendar (plum and cherry) from green tip and bud stage to harvesting time observed in the selected villages 102
4.5 Changes in crop calendar (peach and apricot) from green tip and bud stage to harvesting time observed in the selected villages 103
4.6 Apple production (MT), production area (ha) and productivity (MT/ha) in Kotkhai tehsil 109

4.7 Productivity and climate conditions during different growth phases of apple from year 2007 to year 2017 111
4.8 Pear production (MT), production area (ha) and productivity (MT/ha) in Kotkhai tehsil 112
4.9 Productivity and climate conditions during different growth phases of pear from year 2007 to year 2017 113
4.10 Plum production (MT), production area (ha) and productivity (MT/ha) in Kotkhai tehsil 114
4.11 Productivity and climate conditions during different growth phases of plum from the year 2007 to year 2017 115
4.12 Peach production (MT), production area (ha) and productivity (MT/ha) in Kotkhai tehsil 116
4.13 Productivity and climate conditions during different growth phases of Peach from the year 2007 to year 2017 118
4.14 Apricot production (MT), production area (ha) and productivity (MT/ha) in Kotkhai tehsil 119
4.15 Productivity and climate conditions during different growth phases of apricot from the year 2007 to year 2017 120
4.16 Cherry production (MT), production area (ha) and productivity (MT/ha) in Kotkhai tehsil 121
4.17 Productivity and climate conditions during different growth phases of cherry from the year 2007 to year 2017 123
5.1 Changes in the biodiversity in the field area (Flora) 137
5.2 Changes in the biodiversity in the field area (Fauna) 138

Figures

1.1 Climate Pattern in Himachal Pradesh based on Koppen's classification 11
1.2 Level of analysis 38
1.3 Triangulation method for data analysis 41
2.1 Location of Himachal Pradesh in India 48
2.2 District map of Himachal Pradesh 49
2.3 District map of Shimla 53
2.4 The location of Kotkhai in Shimla (Himachal Pradesh) 54

2.5 Location map of study villages 60
3.1 Mean maximum temperature (in °C) of five decades (1971-2020) 66
3.2 Mean minimum temperature (in °C) of five decades (1971-2020) 66
3.3 Linear regression trendline for maximum and minimum temperature (in °C) 67
3.4 Mean rainfall (in mm) for the five decades (1971-2020) 68
3.5 Mean pre-monsoon rainfall (in mm) for the five decades (1971-2020) 69
3.6 Mean monsoon rainfall (in mm) for the five decades (1971-2020) 69
3.7 Mean post-monsoon rainfall (in mm) for the five decades (1971-2020) 70
3.8 Mean winter rainfall (in mm) for the five decades (1971-2020) 71
3.9 Linear regression trendline for annual mean total rainfall (in mm) 71
3.10 Linear regression trendline for annual mean pre-monsoon rainfall (in mm) 72
3.11 Linear regression trendline for annual mean monsoon rainfall (in mm) 73
3.12 Linear regression trendline for annual mean post-monsoon rainfall (in mm) 73
3.13 Linear regression trendline for annual mean winter rainfall (in mm) 74
3.14 Annual mean snowfall (in cm) for the years 2010-2020 75
3.15 Linear regression trendline for annual mean snowfall (in cm) from 2010 to 2020 75
3.16 Respondents' perception regarding temperature 76
3.17 Respondents' perception regarding overall rainfall 77
3.18 People's perception regarding factors leading to climate change 81
3.19 Factors affecting the perception about climate change 86
4.1 People's perception about the chilling period in the study area 98
4.2 People's Perception regarding flowering time in the study area 98

4.3 People's perception regarding harvesting time in the study area 99
4.4 Scatter chart showing apple productivity (MT/ha) 110
4.5 Scatter chart showing pear productivity (MT/ha) 112
4.6 Scatter chart showing plum productivity (MT/ha) 116
4.7 Scatter chart showing peach productivity (MT/ha) 117
4.8 Scatter chart showing apricot productivity (MT/ha) 119
4.9 Scatter chart showing cherry productivity (MT/ha) 122
5.1 Perception regarding impact of climate change on livelihood among the respondents of the study area 126
5.2 Perception regarding extent of impact of climate change on assets of livelihood among the respondents who observed the effect of climate change on livelihood 126
5.3 Climate change's impact on physical, mental and social health of the people 151

Plates

5.1 Conservation of water from Kangi Nala and rainwater 134
5.2 *Kuphara* in *ghātī* with receding water level 134
5.3 Apple tree during flowering stage 164
5.4 Damaged apple crop 170
5.5 Damaged leaves due to hailstorm 170
5.6 Uprooted tree due to accumulation of snow on the anti-hail nets 171
5.7 Tree with canker, a fungal disease 173
6.1 High density plants 178
6.2 Rooftop gutter 180
6.3 Village Kufar Bag before 2021 185
6.4 Village Kufar Bag in 2021 185
6.5 Anti-hail gun in Deorighat 186
6.6 Rainwater harvesting tank 195

List of Abbreviations

DFID	Department for International Development
DMC	Disaster Management Centre
EEA	European Environment Agency
FAO	Food and Agriculture Organization
GDP	Gross Domestic Product
GP	Gram Panchayat
Ha	Hectare
HDP	High-Density Plant
HP-SAPCC	Himachal Pradesh-State Action Plan on Climate Change
IFAD	International Fund for Agricultural Development
ILO	International Labour Organization
IMD	India Meteorological Department
IPCC	Intergovernmental Panel on Climate Change
IPH	Irrigation and Public Health
IRP	International Recovery Platform
JICA	Japan International Cooperation Agency
LWSS	Lift Water Supply Scheme
MS Excel	Microsoft Excel
MT	Metric Tons
NAPCC	National Action Plan on Climate Change
NWM	National Water Mission
RCC	Reinforced Cement Concrete
RWHS	Rain Water Harvesting Structures
SAARC	South Asian Association for Regional Cooperation

SLF	Sustainable Livelihoods Framework
SPSS	Statistical Package for the Social Sciences
TERI	The Energy and Resources Institute
UNDP	United Nations Development Programme India
UNFCCC	United Nations Framework Convention on Climate Change
UNICEF	United Nations International Children's Emergency Fund
UNISDR	United Nations Office for Disaster Risk Reduction
USGCRP	U.S. Global Change Research Program
WMO	World Meteorological Organization

Glossary

Āpasadārī	Brotherhood and communal bond
Bāna	Ban oak
Bāoṛī/Bāoṛiyān	Step-well/ Step-wells, a source of water
Barāgha	Common leopard
Baṛiyā	Small ball of *Vigna mungo*, the black gram
Bharānī	Workers that sort and fills apple in the boxes
Biśū kā Melā	A social and cultural fair, celebrated on the 9th day of April and is continued for the next two days
Buśaharī topi	Round Himachali cap
Chaiūn	Common morel
Chaśamā/Moka	Aquifer
Chotī	Hill-top
Deva Nītī	The command of their deities
Ḍhātū	A cloth worn on head
Dihāḍī	Daily wages/daily attendance in the field register
Grāma Devatā	Village deity
Iṣṭa Devatā	Local deity
Jātara	A fair celebrated near Mahasu village on the third Tuesday in the month of May.
Jeṭhāgari	Summer rain in the month of *Jeṭha*, i.e., 15th May to 14th June
Kaṛā	Wristband
Kakkaṛ	Barking deer
Kanabāle	Earrings
Kaniṣka	Alder tree
Kasturā	Musk deer

Kathara	Wild goat
Khaḍa	A water catchment area at the bottom of the hill
Kharaśū	Brown oak
Khīna	A function in order to celebrate a special occasion to please and thank the deities
Kuhala	Water channels
Kula Devatā	Clan deity
Kuphara	A natural check dam
Moharū	Green oak
Nāṭī	Traditional dance form in Kotkhai
Pahāḍī	Dialect spoken in the area/people residing in the hills
Parālā	Sunny side of the slope that is south facing slope of the hill
Phīla/Phīlen	Slug/s
Phōṛa	A room with a passage that connects the house to the cowshed
Rīkha	Black bear
Sadrī	A woman's jacket
Sājā	Sankranti
Śīlā	Shady side of the slope that is north facing slope of the hill
Śṭākā maśarūma	Shiitake Mushroom
Tikkara	A huge festival of Kiari panchayat, celebrated after four years to please their local deities
Tuṛāna	Apple picking

1
Introduction

Saving our planet, lifting people out of poverty, advancing economic growth... these are one and the same fight. We must connect the dots between climate change, water scarcity, energy shortages, global health, food security and women's empowerment. Solutions to one problem must be solutions for all.

—Ban Ki-moon

Climate change has become a global threat due to its accelerated rate and detrimental effects on our planet and its inhabitants. Attempts are being made at global and local level to comprehend this phenomenon, the causative agents behind it, its impacts, adaptations required to mitigate it, and policies as well as strategies needed to combat it. The present research is an anthropological study conducted to understand the impact of this global phenomenon on the livelihood of the people at community level.

1.1 Climate change

Climate is the prevalent long-term weather conditions of an area which is determined by latitude, position in relation to oceans or continents, altitude, etc. Weather is day-to-day state of atmosphere. According to Intergovernmental Panel on Climate Change (IPCC, 2012), climate can be simply defined as the average weather, or more thoroughly, as "the statistical description in terms of the mean and variability of relevant quantities over a period of time ranging from months to thousands or millions of years. The classical period for averaging these variables is 30 years, as defined by the World Meteorological Organization. The relevant quantities are most often surface variables such as temperature, precipitation, and wind. Climate in a wider sense is the state, including a statistical description, of the climate system".

The climate of earth has changed throughout the history of the earth. However, never has this change been so rapid in the past. The causes for climate change can be biotic processes, variations in solar radiation received by earth, plate tectonics, volcanic eruptions and anthropogenic activities. As per Intergovernmental Panel on Climate Change (IPCC, 2007a), "climate change refers to a change in the state of the climate that can be identified by changes in the mean and/or the variability of its properties, and that persists for an extended period, typically decades or longer. It refers to any change in climate over time, whether due to natural variability or as a result of human activity". These properties can be variables like precipitation, temperature, humidity, wind, and severe weather events.

United Nations Framework Convention on Climate Change (UNFCCC, 2011) defines climate change as a change of climate that along with the natural climate variability observed over comparable time periods, can be attributed directly or indirectly to the activities of humans which alters the composition of the worldwide atmosphere. The present rate of change of climate has increased manifolds due to anthropogenic activities which have been prominent to such an extent that there has been debate between geologists and environmentalists to call the present times as Holocene or Anthropocene. Conferring to IPCC (2007b), the Earth's climate is changing mainly as a result of the increasing concentration of greenhouse gases in the atmosphere that are emitted from various human activities. Steffen *et al.* (2011), in a study stated that the changes are evident and experimental scientific studies prove that the global environmental change is due to humanity and this has given rise a new geological epoch—the Anthropocene.

This huge human impact on the climate may be a recent phenomenon but the awareness regarding the impact of climate on humans is not new. As mentioned by Dove (2014), Hippocrates wrote an essay called 'Airs, Waters, Places' which claimed a connection between climate, environment and human condition. Montesquieu, a social theorist, viewed a close relationship between social life and climate. He believed that heat made people lethargic and cold air made them vigorous.

The alarming rate of changing climate has led to series of debate regarding its repercussions and has become the burning topic of discussion at national and international platforms. It poses a threat to the settings of human societies as we know them. Research related to climate change is being undertaken within and outside the academic world. International institutions such as IPCC, UNFCCC havc been established to recommend policies and strategies to combat and mitigate climate change. Currently, six IPCC reports have been published, the first in 1990 and the most recent, sixth in August, 2021.

However, the policies formed and implemented by the international bodies have not been effective enough to combat and mitigate the impacts of climate change and to bring the rate of change of climate under control. One of the reasons behind this is that these policies and strategies implemented are not reaching the grassroot level. According to Barnes *et al.* (2013), the climate change mitigation and adaptation projects based solely on natural science are unlikely to succeed as there is no close understanding of the societies in which they are to be employed. Eriksen (2021), suggested that interdisciplinary alliance is required in order to comprehend the full implications of climate change. While climate scientists use a generalised perspective on the planet, anthropologists conduct holistic and in-depth study to access the local realities in order to understand perceptions and response to climate change.

1.2 Anthropology of climate change

Anthropologists study climate change in context of human systems that cause climate change, the ways in which different people perceive and comprehend climate change, its varying effect on people around the world and the various societal mechanisms that are used in adaptation and mitigation. Rendering to Barnes *et al.* (2013), anthropology have three main contributions to study of climate change, which are, its ethnographic insight, historical perspective and holistic view. The ethnographic insight helps in understanding the cultural values and relationships that form the basis of climate related knowledge and interpretation, and the kind of responses to environmental changes. The historical perspective gives the idea about the past society-environment interactions and hence the changes therein with the changing environment. The holistic perspective highlights the numerous cultural, social, economic and political changes that takes place in the society due to climate change.

The traditional strengths of the field of anthropology, participant observation and ability to give insight into the perception and knowledge of the people, the emic viewpoint, can be used to understand the reality of climate change at local level.

1.2.1 History of anthropology of climate change

According to Eriksen (2021), environmental anthropology and the anthropology of energy have been the precursors of the study of climate change. This theoretical approach goes back all the way to the nineteenth century and early studies of material culture, technology, and ecological adaptation. Steward (1955) advocated the study of human ecology where he focused on social and political systems with respect to technology as well as ecology and found that ecological conditions could result in social complexities. White (1949), studied technology and energy and claimed that cultural evolution was directly proportional to the energy being used.

As early as the 1970s, discussions shifted to the study of ecological crises and Bateson (1972), identified three factors that were driving these crises:

- First factor was deleterious side effects of technological progress, for example, production of pesticides.
- Second factor was increasing population that led to depletion of resources.

- Third factor was a set of values and ideas that prompted people to control the environment instead of being a part of it.

He however, did not address climate change clearly. Margaret Mead was the first anthropologist to talk about climate change during a conference about atmosphere (Eriksen, 2021). An early important contribution to study of climate change was made by Rayner and Malone (1998), through their work on knowledge about local livelihoods, political decision-making, and inequality. The path breaking contribution to anthropology of climate change was made by Baer and Singer (2009), who explored the climate change effect on water, nutrition, and the spread of disease. They found that impact of climate change was different for different communities depending on their economic system.

Over the years, considerable body of anthropological literature dealing with different aspects of climate change concerning to vulnerability, resilience, adaptation, mitigation, etc. have appeared.

1.2.2 Importance of anthropological approach in studying climate change

Anthropological approach is important because of the following three assets as mentioned by Eriksen (2021):

- An awareness of variation is essential to all anthropological research. For example, similar policies may not work for developed and developing countries due to difference in infrastructure and socio-economic conditions.
- An insight that any change or policy to be implemented must be managed and made based on the situation at the bottom, i.e., bottom-up approach should be used.
- Comparison in order to generate new ideas.

1.3 Livelihood

Livelihood is defined as the capabilities, assets (both material and social resources) and activities that are required for a means of living and it is sustainable so that it can cope with, and recuperate from stress and shocks, preserve or improve its capabilities and provide livelihoods opportunities for both present and the next generation without undermining the natural resource base (Chambers and Conway, 1991).

Livelihood is a mean to secure the necessities of life, the mode of subsistence which can be viewed and used differently depending on the meaning it has for the people. Livelihood may mean natural resources to some and money to others. According to Scoones (2009), livelihood is a flexible term which can be used along with the topic under study, for instance, rural and urban livelihood or sustainable and resilient livelihood. The variables like land/property, food, knowledge, social, economic and political institutions in a community are intrinsic to the concept of livelihood (Islam and Ryan, 2016).

As per a report prepared by International Recovery Platform (IRP) and United Nations Development Programme India (UNDP-India) (2012), the Sustainable Livelihoods Framework (SLF) has been developed by UK Department for International Development (DFID), which views livelihood as a system and provides a way to comprehend:

- the livelihood assets,
- the context within which a livelihood is established,
- livelihood strategies, and
- factors that make a livelihood vulnerable to shocks and stresses.

The livelihood assets may be perceptible, such as food stores and cash savings, as well as trees, livestock, land and other resources. Assets may also be imperceptible such as privileges as well as access to health services, education, information, and employment prospects. Livelihood assets can also be categorized into the following five groups: human, physical, natural, financial, and social capitals (Table 1.1).

Table 1.1: Types of livelihood assets

Human capital	Household members, active labour, education, health, knowledge and skills
Physical capital	Livestock, equipment, vehicles, houses, irrigation pumps

(*Contd.*)

Natural capital	Access to land, forest, water, grazing, fishing, wild products and biodiversity
Financial capital	Saving/debts, gold/jewellery, income, credit, insurance
Social capital	Kin networks, group membership, socio-political voice and influence

Source: www.fao.org/docrep/0008/a0273e/a0273e04.htm

Livelihoods are formed within social, economic and political contexts which may create opportunity or obstacles for livelihood (Table 1.2). The changing natural environment also shape the livelihood, for instance, the climate and geography, frequency and extent of natural disasters, accessibility of flora and fauna, quality of soil, air and water.

Table 1.2: Livelihood contexts

Social relations	The way in which gender, ethnicity, culture, history, religion and kinship affect the livelihoods of different groups within a community
Social and political organization	Decision-making processes, civic bodies, social rules and norms, democracy, leadership, power and authority, rent-seeking behaviour
Governance	The form and quality of government systems including structure, power, efficiency and effectiveness, rights and representation
Service delivery	The effectiveness and responsiveness of state and private sector agencies engaged in delivery of services such as education, health, water and sanitation
Resource access institutions	The social norms, customs and behaviour (or 'rules of the game') that define people's access to resources
Policy and policy processes	The processes by which policy and legislation is determined and implemented and their effects on people's livelihoods

Source: https://www.unisdr.org/files/16771_16771guidancenoteonrecoveryliveliho.pdf

The livelihood strategies are how people access and use the livelihood assets, within the social, economic, political and environmental contexts. The range of livelihood strategies is huge. An individual may take on several activities to meet the needs. People may engage in activities that contribute to a collective livelihood strategy. Individuals of the same households often take on different tasks to enable the nourishment and growth of the family. The strategies are built to not only use the available assets but also to build resilience to vulnerable situations – shocks like natural disaster, trends like new technology, and seasonal changes due to climate change. People must either use the existing strategies or develop new ones in order to survive in such new vulnerable situations.

1.4 The nexus between climate change and livelihood

Climate change has an influence on all spheres of life of people. It does not impact the environment alone, but also the socio-economic condition of the people. Climate is among the most important determinants of survival and human livelihoods as it determines both how and where human communities live, which foods they can grow, the sources of water for domestic, industrial and other purposes, and how societies and economic activities are organized (IPCC, 2007a).

The nexus between climate (nature) and livelihood (humans) exhibits bidirectional causality, where the climate change has affected the livelihood and the livelihood practices have contributed to climate change. As per the report prepared by IPCC (2014), climate changes have had widespread impacts on human and natural systems which are the result of the human influence on climate system in the form of various anthropogenic activities such as increased greenhouse gases emission due to industrialisation and burning of fossil fuels, etc. Climate change is transforming the lives of people everywhere and these transformations may be swift or slow depending on the way climate change manifests itself.

The climate change and livelihood are related as climate change may lead to natural disaster or abrupt changes in climatic conditions that would affect the income and subsistence of the affected geographies, especially agriculture related mode of subsistence. There is a need to combat climate change as it affects the livelihood of the people. Climate change plays a huge adverse role in places where livelihood is based on natural resources and climate factors such as temperature and precipitation. The increased temperature due to global warming and erratic rainfall pattern reduces agricultural output and affects the quality and yield of crop, thus negatively impacting the livelihood of the people with agricultural mode of subsistence.

The developing countries are mainly at risk as their economies are highly dependent on agriculture. They do not have enough resources or options to combat the damages due to climate change. People in these developing nations will face the brunt of natural disasters, water shortage and hunger due to impact of climate change to a greater extent as they already face social, economic and ecological strains and have inadequate resources that limit their ability to adapt to climate change (Kinyangi *et al.*, 2009).

1.5 The climate and livelihood of the area of study

India, the largest South Asian country, is a developing country with majority of its population dependant on agriculture for livelihood. The climate of the country varies from tropical in the south to temperate and alpine in the Himalayan north, depending on the altitude. According to Hingane *et al.* (1985), based on data from 73 meteorological stations, mean temperature in India has shown a substantial increase in warming amounting to 0.4°C over the last 100-year period. Although climate change is a universal problem, its impacts vary from region to region, country to country, and community to community (Adger *et al.*, 2004). Agrarian communities of the developing countries which have limited access to alternate means of production and are dependent on nature, are affected most by the climate change. Climate change is a major concern for the entire Himalayan region due to its sensitive ecosystem and because the rate of increase in temperature is significantly higher than the global average (IPCC, 2007a).

Himachal Pradesh is a north-west state in India situated in western Himalayas whose livelihood is based on natural resources and sectors like hydel power generation, agriculture, horticulture, forestry and tourism. The on-going climate change may impact the livelihood of the people of Himachal Pradesh as it may alter the distribution and quality of natural resources. As per a report prepared by The Energy and Resources Institute (TERI,

2015), Himachal Pradesh is currently facing challenges related to livelihood due to climate change because:

- It has an agrarian economy with 90 per cent rural population dependent on it for livelihood;
- The state is dependent on rainfall for its economic activities;
- Sustainability of sectors like hydel power plants is dependent on snow and glaciers; and
- The state is under threat of climate induced and other natural disasters.

Figure 1.1 shows the regions in the State under each climatic pattern based on Koppen's classification in a broader sense according to which District Shimla falls under the category of Cwb type for climate, which implies that the climate comprises of sub-tropical monsoon with mild and dry winter and moderate hot summer. But the climate is also dependent on the altitude and as per Economic Survey of 2020-21 (Government of Himachal Pradesh, 2021), the state has been divided into 4 agro-climatic zones – Sub-Tropical, sub-mountain and low hills; Sub Temperate, Sub Humid mid hills; Wet Temperate high hills; Dry Temperate high hills and cold deserts.

The area of study, Kotkhai tehsil falls under the category of wet temperate high hills zone. The livelihood of the people of Kotkhai tehsil primarily depends on horticulture and allied sectors. Two key parameters of climate change that has far reaching consequences on agriculture in general and horticulture in particular are high temperature and erratic rainfall patterns. According to Food and Agriculture Organization (FAO, 2004), these parameters will – reduce crop productivity; cause altitudinal shifts in ecological and agro-economic zones; lead to land degradation; reduce availability of water; rise in sea level and salinization. As Kotkhai belt is famous for apple production, which comprise the main source of livelihood of the people of the area, change in climatic conditions like rainfall and temperature will have an adverse impact on the livelihood of the people.

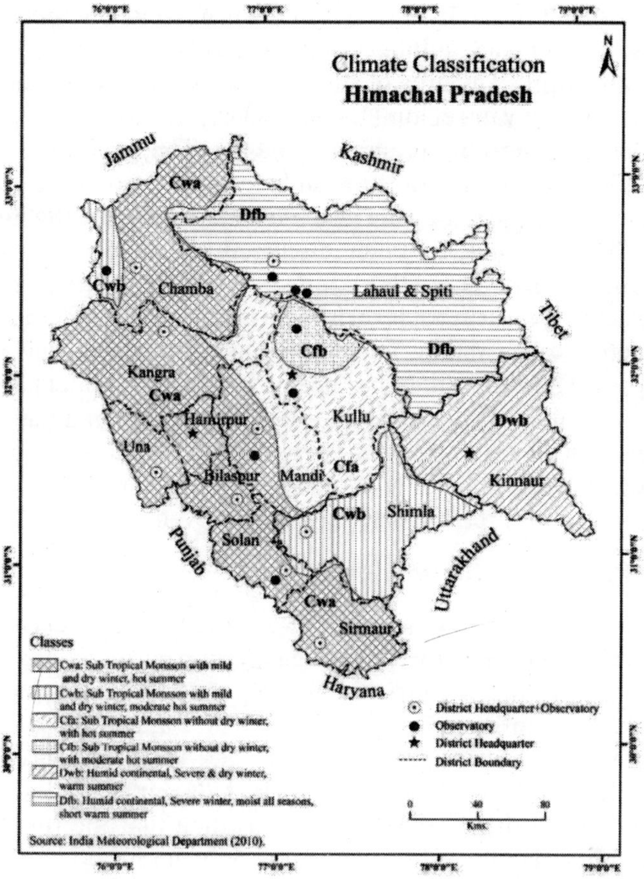

Source: Customised map of map by India Meteorological Department (2010).

Figure 1.1: Climate Pattern in Himachal Pradesh based on Koppen's classification.

1.6 Statement of the problem

According to Intergovernmental Panel on Climate Change (IPCC, 2007a), countries in temperate and tropical Asia are likely to have increased exposure to many extreme events, including the possibility of glacial melts, floods and landslides, rising sea levels, large scale inundation, recession of flat sandy beaches, increased fire risk, water stress, typhoons and tropical storms, and vector-borne diseases. Similar results were found in the 6th

Assessment report by IPCC (2021), which stated that there might be serious impact on India's political economy, especially sector of agriculture. The area of study for present research lies in the inner Himalayan region (1500m-3000m) of Himachal Pradesh which is vulnerable to climate change. The livelihood of the people residing in area is dependent on the horticulture which is directly impacted by the changing climate. While the impacts of climate change are being discussed and studied at national and international level, a knoweledge base related to these impact at the grassroot level is to be taken into consideration for effective policy making and mitigation of climate change and its impact. Thus, this study has been conducted to understand the phenomenon of climate change in the study area and assess its impact on the livelihood of the people, so as to contribute to better understanding and generate data regarding climate change and strategies required to combat it at local level.

1.7 Rationale of the problem

Climate change have managed to get attention of everyone due to its unprecedent rate and negative impact on whole world. One of the researchers had an opportunity to work on the topic, "An anthropological study of cultural adaptation to the climatic conditions and climatic changes: With reference to the Bhil tribe of Block Badgaon, District Udaipur (Rajasthan)" for M.Sc. dissertation course and it was then that the researcher had a first-hand research experience as an anthropologist about climate change and how it was affecting people. This experience prompted the general curiosity in the researcher about the phenomenon of climate change.

After completing the final semester of M.Sc. in Anthropology, the researcher had the chance to visit her ancestral home in Kotkhai where she stayed for four months, June-September of 2016, before pursuing Ph.D. During her stay, the researcher experienced that the weather conditions were different than before which made her think about climate change. The researcher, already being inquisitive regarding climate change and its impact, developed interest in finding out the extent of change in climatic conditions of the area and if it had an impact on the people of the area. During her stay at her ancestral home, the researcher constantly came across conversations related to issues like decline in snowfall and increase in temperature and how these weather events have affected the horticulture-based economy over the years. Being an anthropologist by training, the researcher was drawn towards the various perceptions of the people of the area and decided to conduct an in-depth study in order to understand and comprehend the perception of the people regarding the climate change and how it influenced their livelihood.

1.8 Review of literature

Review of literature is an important step prior to commencement of any research process. Review of literature as defined by Fink (2005), is "a systematic, explicit, and reproducible method for identifying, evaluating, and synthesizing the existing body of completed and recorded work produced by researchers, scholars, and practitioners." It is a search and assessment of already available literature related to the selected topic of research and its critical evaluation. It helps the researcher by providing an intellectual context for one's work. It helps in identifying data, methods and ideas that may be relevant to the topic of research. A thorough review of existing work may help in identifying the gap in the existing knowledge which the researcher might try to fill with her/his own work. It further aids the researcher in putting her/his own work in perspective – is the work being done has something new to add to already existing database of the topic of research, or if it supports or opposes the already existing literature on the topic?

The present study is an attempt to describe climate change and explore its impact on livelihood of the people of Kotkhai tehsil. Climate change has become one of the emerging research topics in the world today as it threatens to destabilize the conditions of human societies as we know them. The unprecedent rate of change of climate has made climate change a topic of debate and research at national and international platform which has led to generation of vast amount of literature. In order to keep the research focussed, for the present study, thematic review of literature has been done keeping in mind the aim of the study. Literature has been reviewed to get an idea about various studies done, concept behind those studies, methodology used and the results achieved. Keeping the findings of the reviews in mind, the present study has been conducted to take into consideration the local perspective and experience of the phenomenon of climate change and its impact on livelihood of the people of the selected area of study.

A thematic division of the review has been done into sections that comprise of work across the globe related to climate change and its impact on livelihood. Considering the anthropological approach of the study, the review of literature is further divided into following five sections, viz.:

- anthropology and climate change;
- climate change trends;
- the impact of climate change on livelihood;
- impact of climate change on agricultural livelihood; and
- adaptations made to combat and mitigate climate change.

1.8.1 Anthropology and climate change

Anthropological insights are making significant contributions to the knowledge and policy making in context to climate change and also addresses the crucial gaps and challenges at the local and global level due to anthropogenic climate change. Crate and Nuttall (2009), emphasize the importance of studying local responses in their edited book, which is a compilation of work of various authors, in a wide range of societies, from Siberia to Papua New Guinea. Hendry (2014), in her work looks at various indigenous cultures as well as knowledge and considers these in relation to western knowledge and science. According to her, lessons can be learned from indigenous people and the way they engage with their environment, for example, Amazonian or Melanesian people who do not alter their ecosystem while harvesting and production and leave a negligible ecological footprint.

Anthropological work identifies the local vulnerabilities and resilience due to climate change. Anthropologists have contributed some significant ethnographic monographs on climate issues such as Barnes' (2014), research on water in the Nile delta and Connor's (2016) work on mining in Australia. Both these studies recognize the global-local linkages and explores how local lives and communities cannot be understood separately without taking into consideration the large-scale processes or global projections of climate change, that result in changed circumstances for future options and constraints.

A ground-breaking contribution was made by Rayner and Malone (1998) to the study of climate change in anthropology. Their work, which comprised of interdisciplinary contributions from around the world, intended to use knowledge about local livelihoods, political decision-making, and disparity to complement the natural science of the IPCC. Ethno-climatological research done by Orlove *et al.* (2000) in the Andes is another pioneering work in anthropology that highlighted how locally rooted knowledge about climate could be of great scientific relevance. According to their study, the farmers predicted temperature change and inter-annual precipitation on the basis of the visibility of the Pleiades star cluster, which in turn depended on El Niño weather events.

The anthropological study of climate change has broadened the horizon of research over the years. The studies are no longer limited to ethnographic accounts, monographs and traditional knowledge base. Crate and Nuttall (2016), sum up the diversification and development of the field by asserting that anthropologists in present times are engaging in research related to resilience, susceptibility, anticipation, adaptation, mitigation, migration, displacement, risk and uncertainty, consumption, and gender.

In the present research, an attempt has been made to understand climate change and its impact by taking into consideration not only the perception of people and local knowledge, but also using the scientific data related to climate conditions available and comparing it with the local knowledge.

1.8.2 Climate change trends

The climate has always been changing; however, it is the increased rate of change of climate that has become a reason of concern as it has started to have an adverse consequence on every domain. According to Weart (2004), anthropogenic climate change is the most striking environmental challenge affecting the earth. Climate change is defined by a number of factors, including: temperature, rainfall, humidity, wind and air pressure, and severe weather events (Schneider *et al.*, 2007; Brown *et al.*, 2012). The climate variables are changing and these changes follow a trend which may be increasing, decreasing or erratic.

1.8.2.1 International studies

Stafford *et al.* (2000), recorded temperature and precipitation from1949 to 1998 to conduct a 50-year trend analysis of the climate variables where it was found that the annual and seasonal mean temperature revealed an increasing trend throughout Alaska, while precipitation showed increase for 3 of the 4 seasons and decrease for summer season. Based on the analysis of precipitation and temperature from 1981-2010 and 1985-2010 respectively, Hayelom *et al.* (2017), observed that the precipitation exhibited an irregular dissemination and temperature presented a general increasing trend in southern Tigray regional state, Ethiopia.

As per World Meteorological Organization (WMO, 2021), the global mean surface temperature was 1.2°C higher for January-October 2020 than the pre-industrial baseline (1850-1900). At this rate, there is a one in five chance of the temperature temporarily exceeding 1.5 °C by 2024. Extreme weather events like floods, heavy rainfall, tropical cyclones, and droughts struck many parts of the world and cases of wildfire were also high in the year 2020.

The recent report by Intergovernmental Panel on Climate Change (IPCC, 2021) states that, each of the last four decades has been consecutively warmer than any decade that preceded it since 1850. Globally, average precipitation over land has likely increased since 1950, with a faster rate of intensification since the 1980s. Global mean sea level has risen by 0.20 m (0.15 to 0.25) between 1901 and 2018. Changes in the land biosphere since 1970 are consistent with global warming. Climate zones have shifted poleward in both hemispheres, and the growing season has on average lengthened by up to two days per decade since the 1950s in the Northern Hemisphere extratropic.

1.8.2.2 National studies

Similarly, various studies related to climate change trends in India have also been conducted. However, due to its vast size and complex geography, the extent and trend of climate change present variations. There are various weather events like droughts, floods, monsoon depressions, heat and cold waves, and snowfall, that affect the climate of India. Dash *et al.* (2007), conducted a comprehensive study to examine climate change in the country. They observed that the atmospheric surface temperature in India had increased by about 1°C during winter and by 1.1°C during post-monsoon. The decrease in the minimum temperature during summer monsoon and increase during post-monsoon months have created an alteration by 0.8°C in the seasonal temperature irregularities, which may further affect the atmospheric circulation. Inter-annual variation in minimum temperature record differences for northern and southern regions of the country. The minimum temperature exhibited a decline in magnitude between 1955 and 1972 followed by a sharp increment thereafter for north India. While the minimum temperature showed a steady increase in south India. Analysis of rainfall amount indicated a decreasing trend during summer monsoon and increasing trend during pre-monsoon and post-monsoon months.

Jain and Kumar (2012), reviewed studies relating to trends in rainfall, rainy days and temperature in India. In this study too, clear and constant picture of rainfall trend did not emerge. Regarding trends in temperature, the mean maximum temperature series demonstrated a rising trend at most of the stations, while a falling trend at some stations. The mean minimum temperature revealed an upward trend in the south, central and western parts of India and a downward trend in some parts of the north and north-eastern India.

De *et al.* (2005) observed that different altitudes in Himachal have witnessed decreasing trend of snowfall at Bhang, Solang, Dhundi and Patseo station for recent 20 years of observations. As per Bhutiyani *et al.* (2007), the annual precipitation, winter and monsoon precipitation over Shimla for the time period 1866-2006 have demonstrated a decreasing trend. Similarly, according to HP-SAPCC (2012), districts of Shimla and Sirmour had a decreasing trend of rainfall for past 25 years and increasing trend for district Kinnaur, Chamba as well as Lahaul and Spiti.

In a meteorological monograph based on the data about the climate change trends at a State level, provided by Indian Meteorological Department, Pune, Rathore *et al.* (2013), studied the long-term changes in precipitation and surface temperature in India from 1951 to 2010. For the State of Himachal Pradesh, data were collected from five stations for temperature, and 36 stations for rainfall. The study pointed that annual average rainfall have declined in Himachal Pradesh with highest significant decrease in monsoon months rainfall in July amongst all the States in India, and a significant decrease in winter months rainfall in January. The overall mean maximum temperature had risen in the most for Himachal Pradesh, with the magnitude of increase being maximum in January and August.

Jaswal *et al.* (2015) analysed seasonal and annual trends in rainfall in Himachal Pradesh, where the annual rainfall showed a decreasing trend. Seasonal rainfall for monsoon and post-monsoon revealed a decreasing trend, an increasing trend in winter and summer for Shimla. The daily heaviest rainfall recorded a decline in areas like Hamirpur, Kotkhai, Nahan and increase in areas located in middle Himalayas like Kalpa.

According to various studies reviewed, temperature has a linear trend of increase while rainfall have shown an inconsistent pattern. There has been an increase in rate of extreme weather events like droughts, floods, heat waves and heavy rainfall. Trends related to surface data like temperature and rainfall have been analysed in present research.

1.8.3 Impact of climate change on livelihood

Livelihood is affected by climate change as there is a link between climate and livelihood practices. People have evolved the ways of earning livelihoods and fulfil their basic need for food, water and shelter along with other services, according to the type of climate in which they live (Kulkarmi and Leary, 2007). The climate has been changing and the deviations that are too abrupt or far from normal can be disruptive for the livelihood that is dependent on natural resources. Effects of climate change tend to be more severe where people depend on climatic factors such as rainfall and temperature for their livelihoods. Gentle and Maraseni (2012) in their study of a remote mountainous district of Nepal found that changing weather patterns had significantly challenged the livelihoods of a community, who were experiencing resource depletion and degradation, food scarcity, lack of basic services, and increasing social disparities.

To assess the impact of climate change on livelihood, one must take into consideration not only what people do in order to make a living is, but also the resources that provided them with the capability to live a satisfactory life (Ellis, 2000). These resources are the assets (human, physical, natural, financial and social capital) which generate livelihood directly or indirectly.

1.8.3.1 Impact of climate change on human capital

Human capital is the asset of livelihood that includes education, household members, health and employment as factors that would affect the way of life of an individual or a community. Climate change has a substantial impact on the human capital livelihood.

1.8.3.1.1 Education

According to Save the Children (2008) and the United Nations International Children's Emergency Fund (UNICEF, 2008), climate change will interfere with the educational provision. Randell and Gray (2019), in their study found that extreme climate changes like excessive heat and precipitation in the tropics could affect the educational attainment. The extreme weather events such as tropical cyclones may damage school buildings, leaving children temporarily unable to attend school, who might even never return to their studies. Households with their income and food security being compromised due to climatic event may not be able to support the education of the children and some may pull children out of school to help earn extra income.

1.8.3.1.2 Household members

Household members are a source of income, due to climate change migration of people in search of livelihood is taking place which may or may not result in economic stability. According to World

Meteorological Organization (WMO, 2021), the climate and weather events have instigated significant population movements and have adversely affected vulnerable people on the move. Asian Development Bank (2012) reported that the impact of climate change manifested as floods, cyclones, droughts, etc. will lead to outmigration of the people residing in low-lying Delta region of Bangladesh. Climate change will cause a significant increase in distress migration in India. According to Salve (2013), migrants leave their homes in search of temporary income sources and are susceptible to exploitation, especially the ones who migrated due to climate-related changes.

1.8.3.1.3 Health

A study performed by Costello *et al.* (2009), illustrated that the climate change is the biggest global health threat of the twenty-first century. As per the report prepared under U.S. Global Change Research Program (USGCRP, 2016), the climate change may impact the physical, mental and community health. Climate change can affect human health in several ways, like, by changing the severity or incidence of health problems and by creating new unexpected health problems or threats in places where they have not previously occurred.

The observation of Dhara *et al.* (2013), recorded that India as a developing country and its high population density, might experience numerous human health effects like water-borne and vector-borne diseases such as malaria, cholera, and dengue because of climate change. Health effects may also include health consequences due to extreme weather events that lead to loss of housing, hunger, and injuries, other than infectious diseases.

1.8.3.1.4 Employment

Climate change has an undeniable influence on employment beyond doubt. According to International Labour Organization (ILO, 2008), more than 1 billion people are working in the agriculture sector, which is the second greatest source of employment worldwide after services and 60 per cent of them belong to China and India together. The extreme and unpredictable weather events caused by climate change will affect the agricultural sector and thus the livelihood of the people associated with it.

1.8.3.2 Impact of climate change on physical capital

Physical capital comprises of livestock, equipment, vehicles, houses, irrigation pumps, etc. Findings of Olsson *et al.* (2014) highlighted that damage to physical assets due to climate events is well documented for settlements often built-in risk-prone floodplains and hillsides susceptible to erosion and landslides. Impacts include houses demolished by flood water and disrupted water and sanitation services. As per Hewitt and Mehta (2012), climate change events like flash floods have severely damaged orchards, roads, and stream embankments in the Himalayas. According to Food and Agriculture Organization (FAO, 2016), climate change impacts livestock directly through heat stress and increased morbidity and mortality, and indirectly through quality and availability of feed and forages, and animal diseases. The United States Global Change Research Programme (USGCRP, 2014), and Transportation Research Board and National Research Council (2008) report that transport, road accessibility, vehicles, etc. can be damaged by the various aspects of climate change.

1.8.3.3 Impact of climate change on natural capital

Natural resources can be used for economic gain to support livelihood which comprise of land or soil, air, water, amongst others. There is a growing agreement that natural resources which act as livelihood assets are closely connected to climate change. In accordance with this statement Xu *et al.* (2009), reported in their study that cascading effects of rising temperatures and loss of ice as well as snow in the Himalayan region were distressing water availability, biodiversity, and were leading to ecosystem boundary shifts like high-elevation ecosystem changes and monsoonal shifts.

1.8.3.3.1 Impact of climate change on water

Increase in temperature, variation in precipitation patterns, and rise in the sea level can affect the water quality and its availability in many regions. A report of IPCC (2007b), projected that by 2050s climate change will lead to decline of available freshwater basins in central, southern, eastern and south-eastern Asia. On a similar note, the International Fund for Agricultural Development (IFAD, 2009) postulated that climate change had affected the quality of water in many areas of the Asia/Pacific region and therefore adequate management was required to guarantee secure supply to the fast-growing populations. Natural water resources in India are being threatened constantly by the impact of climate change, and this threat has culminated in increased water stress.

According to the United Nations Office for Disaster Risk Reduction (UNISDR)/South Asian Association for Regional Cooperation (SAARC), Disaster Management Centre (DMC), (UNISDR/SDMC, 2014), the unprecedented change in climate have severe effect on the hydrological cycle and water resources, thereby leading to drought, flood, scarcity of drinking water, etc. The impact on water resource have further influenced the food security, health, human settlements and on other related areas in the SAARC region. Goyal and Surampalli (2018) stated that changes in precipitation in India will affect water resources, their planning and management in several ways. India's agrarian economy, relies heavily on the monsoon and available water for production is particularly vulnerable to the projected climate changes.

1.8.3.3.2 Impact of climate change on soil

Soil appears to be more vital for modern human societies than ever before to meet the global demands for food and fibre of increasing population. There is a direct linkage between soil and climate change because climate alteration will inevitably affect the nature and distribution of ecosystems in the soil. The soil holds water as well as nutrients, and serve as a growing medium for plants. According to European Environment Agency (EEA, 2015), there are clear signs that soil moisture content is being affected by changes in precipitation patterns and rising temperatures. According to Pareek (2017), India has 9 out of the 12 soil orders that exist in world and 15 agro-climatic zones.

Soil development procedures and properties related to crop production will be affected by climate and will thus influence the livelihoods of millions of people in the country.

1.8.3.4 Impact of climate change on financial capital

The impact of climate change on financial capital is highlighted by the decrease in income, and by poverty. According to Olsson *et al.* (2014) impact of climate change on livelihood could result into loss of profit, unemployment, and poverty. The increased rate of poverty among the people dependent on climatic conditions for livelihood can be connected to the adverse effect of climate change. A plethora of previous studies (Hassan and Nhemachena, 2008; Iwasaki *et al.*, 2009; Alderman, 2010; Jabeen *et al.*, 2010) attributed loss of financial assets to climatic change by losing farm income and jobs. According to Thomas *et al.* (2007), income losses due to weather events mean less money for agricultural needs, school tuition, books and uniforms, and health expenditures throughout the year.

1.8.3.5 Impact of climate change on social capital

Climate change also wear away social and cultural assets. In some contexts, as studied by Buechler (2009), climate changes have disrupted informal social networks of the women, elderly, poorest, and women-headed households by affecting sharing of gifts and deployment of labour in Sonora (Mexico). It was established by Douglas *et al.* (2008) that formal social networks have also been affected by changing climate in Africa. Previous studies by Bell *et al.* (2010) and Green *et al.* (2010) witness that the cultural points of reference are disappearing in cases of indigenous people

The study by Oduniyi (2014) in Mpumalanga province (South Africa), found that climate change affected individuals and groups differently, for instance, the elders, children, tribal group, and low-income population were more sensitive to changing climatic condition than others. The effect of climate change on social capital reflects on networking with other farmers, forming association and cooperatives.

1.8.4 Impact of climate change on agricultural livelihood

Climate change has a very severe negative effect on the agriculture based livelihood. According to Stringer *et al.* (2007), crop yields are declining due to increased parasitic weeds and insect attacks. The impact of climate change is very likely to affect food production at the global, regional, and local level. In every society, farming based livelihoods are issues that are vulnerable to climate change variability. In a study by Saadat and Saiful Islam (2011), it was found that due to the climate change overall access of farmer group to the natural capital, physical capital and financial capital was reduced.

The findings of the study by Vidaurre de la Riva *et al.* (2013) stated that climatic changes for past decades in the study area of Charazani of Andean region (Bolivia) has resulted in extreme weather events, particularly drought, frost, hailstorms along with landslides and increasing cases of fire. The people believed that these climatic events were the main reasons for agricultural loss and had a negative impact on the lives and livelihoods of the people depended on local resources.

In a study of the farming community of Maruf Kebele, (Ethiopia), Belachew and Zuberi (2015) found that the community studied had many negative impacts of the climate change like reduced crop yield, increasing pests/diseases drying up of environment, disappearance of trees as well as plants and soil loss affecting natural plant regeneration. Campbell *et al.* (2016), postulated that climate change will have predominant impacts on crop, livestock and fisheries production, and will intensify the prevalence of crop pests. According to Tuncok (2021), climate change is going to impact four key sectors, that are, water, agriculture, energy, and environment in Central Asia which have adverse implications on water security, food security, energy security, human health, etc.

The productivity of apple was found to be affected by non-availability of water at various critical stages of growth in Shimla district of Himachal Pradesh in a study conducted by Randev (2009). Additionally, productivity and weather parameters have revealed that temperature variations act as a key factor in disturbing the hydrological cycle and intensity of frequency of other parameters.

Climate change have led to lowering of horticultural produce. Hazarika (2013), observed that due to climate change, horticultural crops were showing low production and horticultural crops suffered a yield loss of 10-100 per cent based upon crop and variety. He further noticed that apple production in Himachal Pradesh has shown a decline because the increasing temperature has deteriorated the fruit quality. This may be attributed to lack of chilling hours during winter have led to poor flowering, irregularity in flowering duration, pattern of flowering and poor yield in fruit.

According to Datta (2013), the climate change impact like melting of ice cap have led to reduced chilling effect required for the flowering of many of the horticultural crops in Himalayan regions, therefore, affecting the commercial production of horticulture plants. Also, due to high temperature conditions physiological ailments in horticultural crops will be more pronounced.

The study by Rautela and Karki (2015) found that people of the *Bhotia* tribe inhabiting the northern frontier of Uttarakhand, believed that due to change in climatic conditions like decline in snowfall, increase in temperature and changes in rainfall pattern, productivity of the agricultural fields have declined and have affected the flora and fauna of the area along with the spread of invasive species.

1.8.5 Adaptation made to combat and mitigate climate change

Adaptation is generally 'adjustment in natural or human systems in response to actual or expected climate stimuli or their effects, which moderates harm or exploits beneficial opportunities' (McCarthy *et al.*, 2001). In essence, adaptation is a deliberate change as a reaction to an external stress like adjustments made to changed environmental circumstances (Nelson *et al.*, 2007).

Rendering to Tompkins *et al.* (2010), adaptation to environmental change has always been happening. Societies have adapted to their environments to alleviate risks associated with changes in climatic conditions throughout human history. Rescia *et al.* (2008) observed that this interaction is embodied by cultural landscapes which constitute a witness of the past and present-day associations between society and its environment.

Conferring to Datta (2013), the most effective way to combat climate change is to adopt conservation agriculture, use renewable energy, conservation of water, reforestation, etc. In order to minimize the impact of climate change, present horticultural practices need to be modified. Introducing new varieties of horticultural crops that are tolerant to high temperature, resistant to pests and diseases, give high yield under stress conditions, and do not exhaust the natural resources are the main strategies to combat the impacts of climate change.

As observed by Vidaurre de la Riva *et al.* (2013), in their study in Charazani area of Andean region (Bolivia), the climate change was affecting the livelihood of the people, and several adaptive measures were identified at household, community and external levels in order to cope with the events of climate change. The people used traditional knowledge and contemporary practices in agriculture and are also looking up for non-agriculture related activities to support the subsistence.

The study by Belachew and Zuberi (2015) of the farming community of Maruf Kebele, Central Oromia (Ethiopia), addressed that the farmers diversify their livelihood options as an adaptation to inconsistency in the local weather and adopted a number of steps to modify agricultural practices. The results of their study indicate the relevance of understanding of community perception while formulating policies regarding adaptation strategies to deal with the impacts of climate change.

Le and Vo (2021), in their study found that the people living in the Mekong Delta had four types of livelihood adaptation in response to changing climate. These were changes in the structure of agricultural systems, change of employment sites, relocation and out-migration. The local economic structures and the household livelihood resources had important roles in driving adaptive solutions.

It is important to understand the degree and impact of changing climatic conditions on livelihoods and adaptive capacity at regional and local scales as almost 50 per cent of the earth's natural ecologies face similar challenges (Barnosky *et al.*, 2012).

1.9 Aim and objectives

The present study focuses on the four villages of Kotkhai tehsil viz., Bhawana, Kiari, Kufar Bag and Jashla. These villages were selected on the basis of their relative altitude. The study attempts to assess climate change by analysing the available climatic data for the region, which are the temperature, rainfall and snowfall of the region. The impacts of these changes on the various aspects of livelihood of the people have been explored using a holistic anthropological approach.

The aim of the present research is to study climate change and its impact on the livelihood of the people of Tehsil Kotkhai, District Shimla, Himachal Pradesh. In order to evaluate the impact of climate change on the livelihood of the people of the field area following objectives have been formulated:

- To analyse climatic data (precipitation and temperature) of study area.
- To study the change in climatic conditions and its impact on day-to-day activities in the field area.
- To document crop varieties, crop production, their cropping time and their shifting calendar due to climate change.
- To see the impact of climate change on the socio-economic dimension and livelihood of the people of the field area.
- To analyse effects of climate change on health condition of people.
- To study the adaptations, if any, made by the people due to climate change.
- To examine in field area, policies, if any, formed by the government to combat climate change in the given area.

1.10 Research methodology

The research methodology is a way, a process, to methodically solve the research problem. It explains the various steps taken to conduct the study and the logic behind those steps. It comprises of not only the research methods, i.e., the techniques and methods used for conducting research, but also the explanation behind the particular method or technique being used. Research design is a systematic framework in which particular research is conducted. According to Kothari and Garg (2014), a research design comprises of a clear statement of the research problem, procedures and techniques used to gather data, population to be studied, and methods for analyzing data.

The purpose of research is to answer the questions by using quantitative and qualitative tools and can be of different types – exploratory or formulative, descriptive, diagnostic, and hypothesis-testing research study. The present research is both descriptive and exploratory in nature. It is descriptive in the sense that it describes the phenomenon of climate change by collecting and analyzing the recorded climatic data of the field area. The study is exploratory in nature as it investigates and attempts to comprehend the impact of climate change on the livelihood of the people of the field area.

1.10.1 Universe and unit of study

According to Lavrakas (2008), the universe consists of all study elements that are going to be included in the research study. The exact definition of the universe for every study is different and is set by the research question, which instructs who or what is of interest. The universe of study may be area, individuals and group of people, organizations, or even objects. The universe of study for the present research is Kotkhai tehsil of District Shimla, Himachal Pradesh, India.

The unit of study is the small accessible part of the universe of study that is selected based on certain criteria. The unit of study for the present research are the households of the four villages, viz., Bhawana (near Gumma of Tehsil Kotkhai), and Kiari, Kufar Bag and Jashla of Kotkhai tehsil.

1.10.2 Rationale of the selection of study area

The villages for unit of study were selected after conducting pilot survey from 31st May to 30th June, 2017 during which it was found that the villages at different altitude had varying impact of changes in climate conditions. Pilot survey or pretest is a preliminary study carried in order to assess the feasibility, extent, cost, difficulties, if any, for the research to be conducted. It helps to know if the methodology to be used for the research is applicable in the field area.

The villages are selected on the basis of their relative altitude and are categorized into lower, lower-middle, upper-middle and higher altitude villages as per the heights measured by the researcher during fieldwork from the approximated centre of the villages with the help of the mobile application, GPS Tools 3.0.2.4 (Designed and Developed by Virtual Maze). The village Jashla which lies at an altitude of around 2438 m is categorized as the high-altitude village, as area that reaches at least 2400 m from sea level is considered as high altitude (Resource Library, National Geographic). In comparison to Jashla, Bhawana (1825 m) lies below 2000 m and has been considered as a lower altitude village (Kiari (2224 m) is considered as lower-middle altitude village, and Kufar Bag (2362 m) is above Kiari, therefore categorized under upper-middle altitude village, for the purpose of present study.

1.10.3 Rapport establishment

Rapport establishment is very important in the data collection. As fieldwork is the core part of anthropology, the researcher goes to field area and has to convince the respondents to invest their busy time to give some information. The researcher has to develop alliance in the field. It is vital to explain the purpose of the study in order to collect the factual data from the respondents. For the present study, rapport was first established with the key informants during the first field visit in every village with the help of mutual acquaintances. The key informants then further introduced the researcher to other informants or respondents. An attempt to establish a healthy and good rapport with the respondents was made. As one of the researchers also belonged to the same area, the people of the field area were polite and eager to help in research, thus facilitating rapport establishment. The overt nature of research and asking for consent before conducting the research also helped in building the confidence and trust in the process of research.

1.10.4 Sampling

Sampling is the process of choosing a representative part of a population in order to determine parameters or features of the whole population. A sample is a sub-section of population that represents the whole population. By sampling we study only a part of universe, instead of whole universe. The conclusion is then drawn on the basis of that part of universe for the entire universe. Sampling helps to save effort, time and resources by limiting the size of data and making it easier to collect, analyse and interpret the data.

After the selection of the field area, that are the villages Bhawana, Kiari, Kufar Bag and Jashla, the sample for conducting the research was selected, which comprised of the respondents from the alternate households in every village. Certain inclusion and exclusion criteria were applied before choosing the respondent.

1.10.4.1 Sampling technique

There are two main types of sampling techniques:

1. **Probability Sampling:** Each unit of population has equal chance to be selected as a sample.
2. **Non-Probability Sampling:** Specific and limited units of population have a chance to be selected as a sample based on some predetermined reason of the research.

For the present research, purposive and snowball non-probability sampling techniques were employed. Purposive sampling technique is selective in nature as sample is selected on the basis of certain characteristics that would fulfil the objectives of the study. Purposive sampling is done with the purpose in mind. The researcher would be seeking for one or more specific predefined groups. Snowball sampling technique is a technique in which the researchers make initial contact with the small group of people who are relevant to the research topic, for example, key informants, and uses these to establish contact with the other respondents.

The respondents above 30 years of age were selected by purposive sampling as climate change is inferred from observation of at least three decades. Rapport with the key informants in all four villages was established and with the help of these informants, more respondents were reached out to by using snowball sampling as the key informants had an idea about all the households and whether the people were permanent residents of the villages or not.

1.10.4.2 Inclusion criteria

- The respondent must be a long-term resident of the village, minimum 30 years or should have land in the village.
- The respondent must be above 30 years of age.

1.10.4.3 Exclusion criteria

- People below 30 years of age.
- Temporary labourers and workers.
- People with memory issues.

1.10.4.4 Sample size

Table 1.3 depicts the criteria for sample size selection for the present study. A total of 97 households were selected after inclusion and exclusion criteria were applied.

Table 1.3: Criteria for sample size selection

	Selected villages of Kotkhai tehsil	**Altitude (Approximated centre of village from the sea level)**	**Number of house-holds**	**Expected sample size (Alternate house-holds)**	**Actual sample size**
1.	Bhawana	Low altitude (1825 m)	34	17	17
2.	Kiari	Lower-middle altitude (2224 m)	66	33	30
3.	Kufar Bag	Upper-middle altitude (2362 m)	28	14	14
4.	Jashla	High altitude (2438 m)	127	63	36
	Sample Size		**255**	**127**	**97**

Source: Field data.

1.10.5 Details of field visits

Pilot survey was conducted from 31st May to 30th June, 2017, while data were collected in different phases from May, 2018 to April, 2021.

- First field visit to the villages was conducted in the months May-July, 2018 after getting ethical clearance.
- A second field visit for fieldwork was done in the months of February and March, 2019.
- Third visit to field for data collection was done in the May-June, 2019.
- Fourth field visit was conducted in the months of October-November, 2019.
- Fifth field visit for data collection was done in the months of August-September, 2020.
- Last field visit was done in April 2021.

1.10.6 Ethical considerations

Ethical considerations are the most important aspect of any research. It is the duty of the researchers to make sure that their research does not harm the participants or respondents in any way. Ethical clearance was taken before the collection of data from Institutional Ethics Committee, Panjab University, Chandigarh.

As the present study revolves around the livelihood practices and includes financial information as part of the asset of livelihood, the researcher has made an attempt to not to intrude into matters that may be considered private by the respondents. Only the data that the respondents were willing to share, have been collected and the respondents have not been led or pressurized in any way in order to gather data.

The respondents were explained about the research being conducted, its aim and objectives. Informed consent was taken prior to commencement of the research. As the research brings out the opinion and perception of people related to various aspects, the respondents were assured that no harm would come to them at community or government level. The confidentiality and anonymity of the respondents have been maintained.

1.10.7 Data collection

The most important aspect of any research is the data collection. A good research output is based on good data collection. Participant observation, fieldwork, emic approach and holistic approach to gather first hand observation are the characteristic features of data collection in anthropological research.

Data can be divided into primary data and secondary data. Primary data are collected from the primary source of information, while data collected from secondary source of information are called the secondary data.

The present research is based on qualitative and quantitative data. The qualitative data comprises of the primary data that have been collected from the primary source of information with the help of interview schedule, interviews, case studies, focus group discussion, quasi-participant observation and photography. The quantitative data comprises of secondary data such as climate data – temperature, rainfall and snowfall are based on the secondary source of information like India Meteorological Department (IMD).

1.10.7.1 Primary data collection

The methods used to collect primary data, which was qualitative in nature, have been explained as follows:

1.10.7.1.1 Interview schedule

It comprises of a set of structured questions that are recorded by the researchers themselves. The main advantage of this method of data collection is that the respondent can be explained the questions or research problem and therefore making it easy for them to answer. Also, the respondent is able to express or answer clearly thus helping in better understanding for the researcher. For the present research, an open-ended interview schedule was prepared to gain insight into the perception of the respondents regarding climate change and its effect on the livelihood. The interviews schedule was structured and comprised of a list of predetermined questions and themes to be covered, however the sequence of the questions was flexible depending on the respondent in order to avoid breaking the flow of the answers.

1.10.7.1.2 Interview

Interviews are a form of human communication in which knowledge evolves through dialogue, with the aim of obtaining qualitative description by the respondent (Kvale, 1996). As Gray (2004) has pointed out, interviews are a favoured approach when respondents have difficulties with written language, such as the old respondents in this research. While Willis (2006) mentioned that one of the advantages of interviews is that they generate a wide range of data quickly, allowing the researcher to clarify issues.

The researcher has used the in-depth interview, key informant interview and group interview for collecting the data during field work with the help of an interview guide.

- **In-depth interview:** It is an intensive individual interview technique which explores the perspective of the individual regarding a particular event or phenomenon. In the present study, the heads of selected households were interviewed regarding their perception about climate change and its impact on their lives.
- **Key informant intcrview:** These are in-depth interviews that are conducted by selecting a respondent who has an overall idea about the community, which is the village in the present study.
- **Group interview:** It comprises of an interview of two or more people. The perception of the eligible family members of the selected households are taken into account with the help of group interviews.

1.10.7.1.3 Case study

According to Yin (2003),

> *"A case study is an empirical inquiry that investigates a contemporary phenomenon within its real-life context, especially when the boundaries between phenomenon and context are not clearly evident. The case study inquiry copes with the technically distinctive situation in which there will be many more variables of interest than data points, and as one result relies on multiple*

> *sources of evidence, with data needing to converge in a triangulating fashion, and as another result benefits from the prior development of theoretical propositions to guide data collection and analysis".*

The case study method allows to capture the complete real-life events and their significant features which provides a broader view of the particular phenomenon which is climate change and its impact in this research.

Case study is an intensive study of any subject which may be an event or phenomenon, a person, community, or even an entire culture. Burns (2000) has given six types of case studies:

- **Historical case studies:** These trace the development of an organization or system over a period of time.
- **Observational case studies:** These are based on observation of an event or a person.
- **Oral history case studies:** These are the first-person narratives collected by using extensive interviewing techniques.
- **Situational case studies:** It is based on a particular event.
- **Clinical case studies:** It aims to understand an individual in depth.
- **Multi-case studies:** This is basically analysis of multiple case studies to see the differences or similarities in outcomes.

For the present study individual case studies have been recorded related to the perception of the respondent about the event of climate change and its impact on the livelihood of the people and falls under the category of oral history case studies. As the perception regarding an event, i.e., climate change is studied, the case studies also fall under the category of situational case study.

1.10.7.1.4 Focus group discussion

It is a method of data collection in which people with similar background or experiences come together to share their views on a particular topic of interest. The groups normally comprise of 8-10 people and minimum 4 people. It can be used as a supplement to the data from the in-depth interviews. A particular advantage of focus group discussion is to bring up issues and ideas that would not have emerged without the group interaction (Darlington and Scott, 2002; Tacchi *et al.*, 2003).

For the present study, five focus group discussion were conducted, of which four were conducted at the selected villages, and one focus group discussion was conducted with a group of five women of the selected villages. The researcher used a semi-structured approach with predetermined questions and themes to be covered which were used to initiate discussion on various aspects under study. Each focus group discussion began with an introduction to anthropology and the topic of research, after which participants were asked for consent and permission to record their conversation.

1.10.7.1.5 Quasi-participant observation

Quasi participant observation method includes the qualities of both participant and non-participant observation methods. This type of observation method involving limited amount of participation in the community activities. It is difficult for the observer and the observed to maintain the quality of complete participation and non-participation. In quasi-participant observation, the investigator can participate at times and observe from outside the other times.

For the present study, quasi-participant observation technique has been used as the researcher could only stay in the field for certain period of time and in certain months at a time.

1.10.7.1.6 Photography

Photography as a part of visual anthropology can be used in qualitative research for exploratory purposes. Visual anthropology can be defined as pictorial and visual communication, which consists of anything "made to be seen" (Ruby, 1996). In a research, photograph can often reveal more than words and may provide a better understanding. Visual data can support research by validating the data available from the interviews. Adding a visual means of data collection like photography can enhance fieldwork, since no one method of data collection can adequately record and explain all aspects of a phenomenon (Asch and Asch, 2003).

Photography has been used as a visual representation of the impact or result of the changing climate on livelihood of the people of the field area.

1.10.7.2 Secondary data collection

The following sources were consulted for the secondary data collection:

1. Departmental library, Department of Anthropology, Panjab University, Chandigarh.
2. Meteorological Centre, India Meteorological Department (IMD), Bibra House, Cliffend Estate, Shimla (Himachal Pradesh).
3. Department of Horticulture, Chotta Shimla, Navbahar, Shimla (Himachal Pradesh).
4. Satyanand Stokes Library, Dr. Yashwant Singh Parmar University of Horticulture and Forestry, Nauni, Solan (Himachal Pradesh).
5. Tehsildar's office in Kotkhai.
6. Research article based on international and national studies from different sources.

1.10.8 Data analysis

The next step after data collection in research is data analysis. According to Miles and Huberman (1994), qualitative research generates words and quantitative research generates numbers. The present research comprises of qualitative and quantitative data, therefore, both qualitative and quantitative methods of data analysis have been used. Thematic representation of data has been done. The various objectives have been dealt under specific themes.

1.10.8.1 Qualitative data analysis

For the analyses of the qualitative data collected, an emic approach has been used and micro analyses (Figure 1.2) of case studies is done with the help of narratives which have been transcribed using thc system of transliteration (Appendix).

An emic view is the perspective, belief, perception and understanding of the affected person or population and comprises of the knowledge at the regional and local level. Views may differ from person to person or place to place, for instance, degradation is defined in different ways depending on the context or the place of occurrence (Huber-Sannwald *et al.*, 2006). The approach of quasi-participant observation by being there and recording the emic view of the people has been used. As one of the researchers belongs to the same area as the universe of study, the application of the emic approach was made convenient even when the respondent could not spend the entire time in the field during the research.

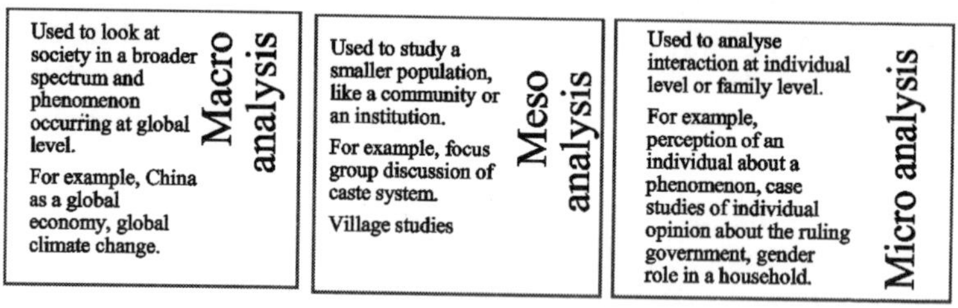

Source: Modified form of https://www.coursehero.com/sg/introduction-to-sociology/macro-level-meso-level-and-micro-level-analysis/

Figure 1.2: Level of analysis.

Narratives from interviews and focus group discussion have been used to analyse the data along with the narratives from micro analysis of case studies. Photographs have also been used for analysis purpose.

1.10.8.2 Quantitative data analysis

Microsoft Excel (MS Excel) has been used for descriptive statistical analysis to analyse the quantitative data gathered from horticulture and meteorological departments as well as to interpret the data as graphs, tables and figures. Statistical Package for the Social Sciences, version 16 (SPSS 16 Incc.) has been used to assess linear regression to analyse the trends in climatic parameter, i.e., temperature, rainfall and snowfall.

The secondary data provided by the Meteorological Centre, Indian Meteorological Department (IMD), Shimla, have been divided into different sets for analysis. The mean maximum and mean minimum temperature have been divided into monthly (decadal), and decade groups. The linear regression has been applied for finding the annual trend of temperature, with mean maximum and mean minimum temperature as dependent variables and years (1971-2020) as independent variable.

The rainfall data provided have been stratified into monthly (decadal), decade, pre-monsoon, monsoon, post-monsoon, and winter rainfall groups. Linear regression has been applied to determine the trend of annual total rainfall, annual pre-monsoon rainfall, annual monsoon rainfall, annual post-monsoon rainfall, and annual winter rainfall from year 1971 to 2020.

In the present research, months of March, April and May are taken as the pre-monsoon months; June, July and August as the monsoon months; September, October and November as the post-monsoon months; December, January and February as the winter months.

The snowfall data have been computed on yearly basis for descriptive analysis as well as linear regression for the years 2010 to 2020. The data provided by the Department of Horticulture has been used to evaluate annual productivity by using the following formula:

Productivity= Production (MT)/Area of Production (Ha)

1.10.8.2.1 Mean

It is the statistic of central tendency used for numeric continuous data. Mean is the average calculated by taking into consideration every observation. For the present study arithmetic mean has been used for temperature, rainfall and snowfall. Arithmetic mean in simple term is 'average value'. It is "the total sum of all the scores divided by the number of scores" (Ahuja, 2007). The mathematical formula for mean is:

$$\underline{\mathrm{x}} = \frac{x_1 + x_2 + x_3 ----- x_n}{N}$$

where x is the mean, $x_1 + x_2 + x_3$--- are the given units in observations and N is the total number of units added together.

1.10.8.2.2 Percentage

A percentage (Latin for per centum which means 'by a hundred') is a number or ratio expressed as a fraction of 100. It is often denoted using the per cent sign, ' per cent'. The per cent value is computed by multiplying the numeric value of the ratio by 100. For example, if 10 respondents out of 50 noticed the difference in climate, the percentage of respondents who observed a difference will be,

$$\text{Percentage} = (10/50) \times 100\%$$

1.10.8.2.3 Regression analysis

It is a statistical technique which is used to study the cause-and-effect relationship between one dependent variable (response) and one or more independent (predictor) variables (Sarma, 2001). Regression analysis can be simple or multiple linear regression. For the present study simple linear regression analysis has been used.

Simple linear regression is applied to find the association between two continuous variables (dependent and independent). Linear regression is used to predict a linear relationship between the input variables (x) and the single output variable (y) (https://machinelearning mastery.com/linear-regression-for-machine-learning/).

A simple linear regression model is a mathematical equation of the type

$$y= b0 + b1x + \varepsilon,$$

Where b0 and b1 are constants to be determined from the data and ε is called the random error component. This is also known as a bivariate regression model. b0 is the regression coefficient called slope of the fitted line and b1 is called the intercept. The adequacy of the linear model is judged with the help of a value called R2 value. It is the square of correlation coefficient and lies between 0 and 1. Also, if the regression coefficient has a positive sign, it means that x is positively correlated with y (Sarma, 2001).

1.10.8.3 Triangulation method of data analysis

Triangulation is a method to analyse results of the same study using different methods of data collection. It is used in order to enhance validity of the data collected, to create a more detailed picture of a research problem, and to cross-examine different ways of understanding a research problem. In most of the cases, it helps to validate research findings by examining that different methods or observer of the same phenomenon produce the same results (Nightingale, 2020).

Triangulation method for data analysis has been used for present study to validate qualitative and quantitative data collected by comparing the results of both type of data with each other. Also, as the researcher's place of origin is same as the universe of study, there was a presumption regarding climate change and its impact on livelihood of the people. In order to avoid dilemma like study being biased and leading, the pre-conceived notions like the increase in temperature and decline in snowfall has been cross-checked with the climate data provided by the Meteorological Department (IMD), Shimla as well as with the data collected by qualitative methods. Researcher's perception regarding declining horticulture practices have been cross-examined with the data collected from Department of Horticulture, Shimla and the data provided by the respondents of the field area. Interviews and discussions were held, without interrupting the respondents or leading them towards a certain answer. The personal views and biases were kept under constant check during data collection. However, during data analysis, the perceptions of the researcher were also kept in mind and compared with the results of qualitative and quantitative data. Figure 1.3 is a representation of how triangulation of data has been done for analysis of data.

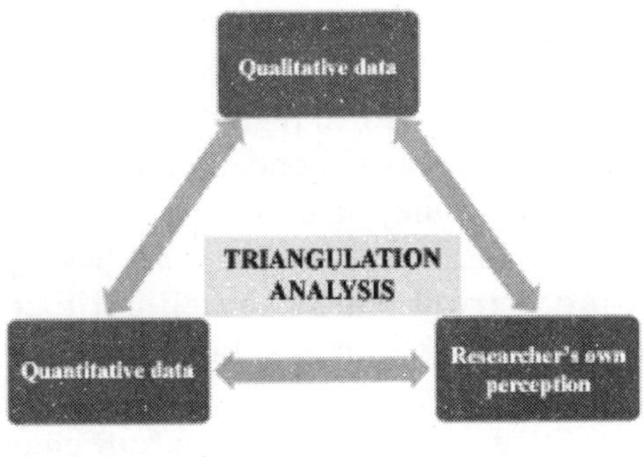

Source: Researcher's concept

Figure 1.3: Triangulation method of data analysis.

1.10.9 Problems encountered

Every step of research journey was challenging and prone to changes. The plan of action with which the researcher embarks the journey of research sometimes changes drastically in the natural settings. The main challenge and problem encountered during the study was unavailability of respondents during field visits as most of the people lived in urban areas and only visited the village during winter vacation or during harvesting of the crop. Due to this, the actual sample size turned out to be lower than the expected sample size.

- The researcher initially wished to conduct interview of respondents of two subsequent generations, however, it was not feasible for all the selected households due to absence of the younger generation respondents, as they were busy with the education schedule of their children in the city.
- It was observed during the fieldwork that most of the households were either vacant as people resided in urban areas or only the elder generation resided there. The land was looked after by the elderly with the help of labour.
- Some of the elderly people could not recall the events from the past thirty years due to poor memory and the labourers working there were not permanent, therefore could not provide the required data.
- It was difficult to access the villages Kufar Bag and Jashla during the winter season due to poor connectivity around snowfall.
- The multiple field visits brought into light difference in the opinion of the people regarding the impact of climate change which mostly depended on the kind of winter or monsoon season they had.
- During the field visit in the year 2018, people were much more troubled by the climatic conditions as there was negligible snowfall during the winter. While when asked about the same during the field visit in the year 2019, because there was sufficient snowfall in winter that year, people did not find climate conditions

problematic. Therefore, inconsistency in data provided by the respondents was also a problem encountered during the study.

- Problem was faced in acquiring the information regarding financial status and land holding as the respondents were reluctant in providing such details in general. Due to the ongoing issue of encroachment of government land for personal use, where people have illegally acquired government land, be it forest area or grassland, for the purpose of horticulture use, the people feared legal action and hence, did not prefer to give the complete details regarding their land holdings.
- Complete data regarding the climate of Kotkhai were not available. However, on comparing, it was found that the climate data for Shimla and Kotkhai were similar, therefore, the data of Shimla have also been used for the purpose of analysis.

1.11 Significance of the study

The universal dimension of climate change is undeniable, but it is also essential to show in what ways climate change is always local in its implications. A holistic approach is required to evaluate climate change and its impact in ecological, social, cultural and political context. There are abundant researches based on natural science related to climate change, however, the findings have not been very successful in implementing policies and mitigation strategies to combat the impact of this global threat. The reason being that there are variations at regional and community level. 'One-size-fits-all' approach is to be abandoned, while formulating policies and strategies to tackle the issue of climate change.

Anthropology provides a view from within and from below which is essential for understanding the climate and society relationship. The present study is an anthropological study of climate change and its impact on the livelihood of the people of Tehsil Kotkhai of District Shimla, Himachal Pradesh. The study is expected to give information about major climate change, impacts and livelihood strategies that people are using at local levels. The study may contribute to improve and facilitate exchange of climate knowledge and data among field experts, researchers and policy makers at grassroot level. The result of the present study may also be used to validate the efforts of the local people to deal with climate change, which can be used as an input while formulating policies and strategies at regional and national level.

The result of the study will provide data about people's perception regarding the climate change and its impact, actions taken by the people at local level to combat and adapt to climate change, mitigation strategies and policies to be developed for their effective implementation. Also, being an anthropological study, it is one of its kind in the present field area, therefore, the findings can be used for further research and development activities related to the study area in particular and the region or nation at large.

1.12 Limitations of the study

The study of climate change impact on livelihood, which is horticulture based, is a long-term study and would require plenty of resources, time and scientific instruments and methods.

This study was constrained in a variety of ways:

- No scientific tools were used during the research as the researcher did not specialize in the field of horticulture and climate.
- Only the variable like temperature, rainfall and snowfall were used to study the impact of climate change.
- The snowfall data were available only for 11 years for the field area.
- Data for fruit production and area for Kotkhai tehsil were available only for the years 2007-2017.

1.13 Chapterization

The book comprises of seven chapters in total. The first chapter Introduction, comprises of an introduction to – climate change; anthropology of climate change; livelihood; the nexus between climate change and livelihood; the climate and livelihood pattern of the area of study; statement of the problem; rationale of the problem; review of literature; aim and objectives; research methodology used to conduct the present study: the universe and unit of study, sampling, methods of data collection, data analysis and problems encountered; significance of the study; and chapterization.

Second chapter is dedicated to Area and People and comprises of brief description about India, Himachal Pradesh and Shimla district along with the universe of study, i.e., Kotkhai tehsil. Demography, economy, religion, language, rituals, customs and festivals of the area of study are also discussed in this chapter.

The third chapter comprises of analysis of climatic data for the past five decades, i.e., from the year 1971 to 2020, and people's perception about climate change and its impact on their day to day lives. The chapter covers the first two objectives of the present study, i.e., to analyse climatic data (precipitation and temperature) of study area, and to study the change in climatic conditions and its impact on day-to-day activities in the field area.

Fourth chapter documents detailed description and impact of climate change on the crop varieties, crop calendar and crop production and comprises of the third objective, i.e., to document crop varieties, crop production, their cropping time and their shifting calendar due to climate change. The chapter includes both, people's perception and the data provided by the Department of Horticulture, Shimla, Himachal Pradesh.

Chapter five analyses the impact of climate change on the livelihood of the people of Kotkhai tehsil by taking into consideration the perception of people about the climate change's impact on different assets of livelihood. The chapter deals with the fourth and fifth objective, to see the impact of climate change on the socio-economic dimension and livelihood of the people of the field area, and to analyse effects of climate change on health condition of people.

The sixth chapter discusses the strategies and efforts to combat and mitigate climate change in the study area and the last two objectives, which are, to study the adaptations, if any, made by the people due to climate change, and to examine in field area, policies, if any, formed by the government to combat climate change in the given area.

The last chapter, chapter seven, comprises of the summary and conclusion of the present research which is followed by references cited and appendices.

2
Area and People

The present research has been conducted in selected villages, viz., Bhawana, Kiari, Kufar Bag and Jashla, of Kotkhai tehsil of District Shimla, Himachal Pradesh, India. In this chapter, the researcher has discussed the various aspects related to the area under study and its people. A brief description about India, followed by details about Himachal Pradesh and Shimla has been given. After this, the area and people of Kotkhai have been discussed and finally the details regarding location and demography of the selected villages have been given.

2.1 India

India is the seventh largest nation in the world in terms of area and largest in South Asia. Located north of the equator between latitudes 8°4′ north to 37°6′ north and longitudes 68°7′ east to 97°25′ east longitude, India is a country known for its cultural diversity. The country with second largest population in the world is bordered by Afghanistan and Pakistan in the north-west; Myanmar and Bangladesh in the east; China, Nepal and Bhutan in the north and north-east; and has the Arabian Sea in the south-west, Indian Ocean in the south, and the Bay of Bengal in the south-east. The archipelago of Lakshadweep, and Andaman and Nicobar Islands, which are the Union Territory of India lie in the Arabian Sea and the Bay of Bengal respectively. There are 28 states and nine union territories in India.

The mainland on India can be divided into four regions, viz., the great mountain zone, plains of the Ganga and the Indus, the desert region and the southern peninsula. The geological regions may be grouped into three regions: the Himalayas and their associated group of mountains, the Indo-Gangetic Plain and the Peninsular Shield.

The climate of India may be largely described as tropical monsoon type and the Indian Meteorological Department (IMD) have designated four official seasons, that are, winter, summer, monsoon and post-monsoon season. However, as the Himalayan states like Himachal Pradesh are more temperate, they experience autumn and spring. Other than the four seasons designated by IMD, traditionally, there are six seasons: *Vasanta* (the spring), *Grisma* (summer), *Varsa* (monsoon), *Sharad* (autumn), *Hemanta* (pre-winter), *Shishira* (winter). These six seasons are two month long and are based on astronomical division of 12 months into 6 parts. The climate is affected by two seasonal winds: the south-west monsoon (which brings most of the rainfall in the country) and the north-east monsoon (Government of India, 2020).

India has total population of 1.2 billion. According to Census (2011), 54.6 per cent of the population is engaged in agriculture and allied activities.

2.2 Himachal Pradesh

Himachal Pradesh, which literally means snow abode, is also known by the name of "Dev Bhumi: the abode of gods and goddesses". Himachal Pradesh is largely a mountainous State and is located in North-West India and is located between latitudes 30°22′N to 33°12′N and longitudes 75°47′E to 79°04′E. The state which ranks 18th in terms of area and 21st in terms of population among states in India is bordered by Jammu & Kashmir in the north, Punjab and Chandigarh in the west, Haryana in the south-west, Uttarakhand in the south-east, and by the Tibet Autonomous Region in the east (Figure 2.1).

Himachal Pradesh is also called the "Fruit Bowl" of the country. The state has favourable climatic conditions to produce a variety of fruits such as apple (*Pyrus malus*), pear (*Pyrus communis*), peach (*Prunus persica*), plum (*Prunus domestica*), apricot (*Prunus armeniaca*), mango (*Mangifera indica*), litchi (*Litchi chinensis*), guava (*Psidium guajava*), strawberry (*Fragaria *ananassa*) and citrus fruits. According to Sharma (2011), Himachal is known as one of the most dynamic hill states of India and scores significantly high on indicators of human development.

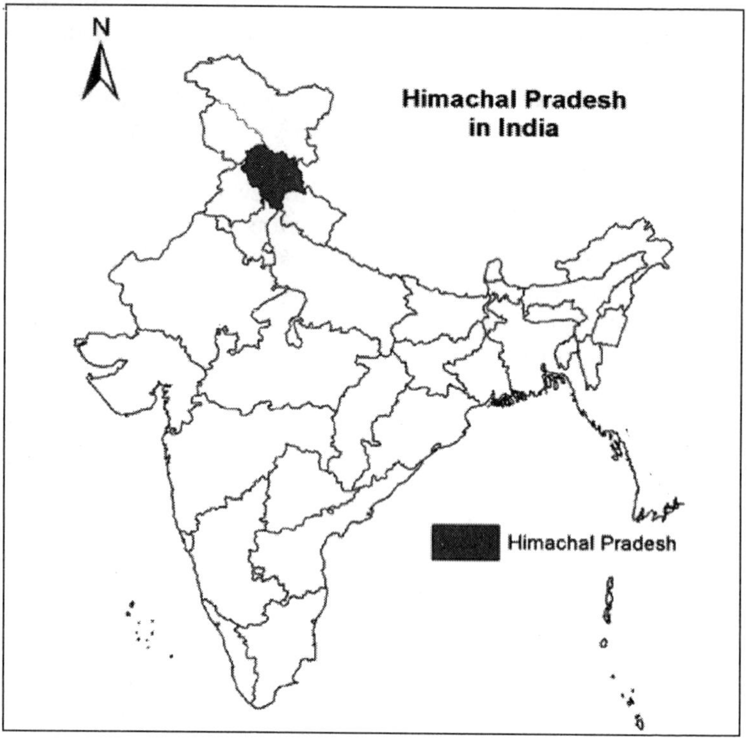

Source: Customised form of map from Census of India.

Figure 2.1: Location of Himachal Pradesh in India.

2.2.1 Topography

Himachal Pradesh lies in the western Himalayas, thus, present an intricate pattern of valleys, mountain ranges and hills. It is a state with mountainous topography with altitude varying from 350 meter to 4500 meter above the mean sea level. It has an area of 55,673 square kilometers that can be divided into three zones:

- Outer Himalayas or the Shivaliks (350 m-1500 m).
- Inner Himalayas or Mid-Mountains (1500 m-3000 m): The field area for the present study lies in this zone.

- The Greater Himalayas or Alpine Zone (3000 m-4500 m).

2.2.2 Administrative units

Himachal Pradesh acquired its present disposition in phases over time after independence. The State of Himachal Pradesh is divided into 12 districts as shown in Figure 2.2, which are grouped into three divisions, Shimla, Kangra and Mandi. The districts are further divided into 69 sub-divisions, 78 blocks and 145 Tehsils.

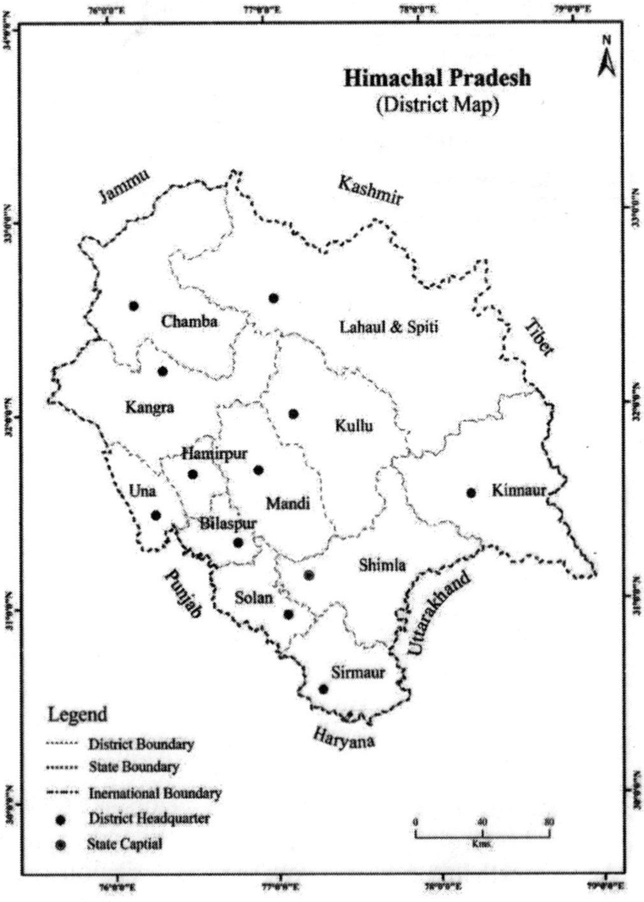

Source: Customised Map.

Figure 2.2: District map of Himachal Pradesh.

2.2.3 Demography

As per the Census detail of year 2011 shown in Table 2.1, around 90 per cent of population of Himachal Pradesh inhabits the rural areas of the State. These villages are sparsely distributed across the State having population density as low as one person per square kilometer in the remote and tribal area of Lahaul & Spiti (District Disaster Management Authority Shimla, 2012).

Table 2.1: Statistical detail of Himachal Pradesh

1	Total Districts		12
2	State Capital		Shimla
3	% Area to the total area of India		1.7
4	Total Area (sq. km)		55,673
5	Total Population		68,64,602
6	Rural Population		61,76,050
7	Urban Population		6,88,552
8	Area under Forest (sq. km)		37,033
9	Population	Male	34,81,873
		Female	33,82,729
		Scheduled Caste	17,29,252
		Scheduled Tribe	3,92,126
10	Sex Ratio		974
11	Literacy Rate		83.78%
12	No. of Inhabited Villages		17,495
13	No. of Towns		59

Source: Census of India, 2011.

2.2.4 Economy

The economy of the state is dependent on sectors like hydel power generation, horticulture, agriculture, forestry and tourism. The economic growth in the State is governed by agriculture activities. Though the economy has shown a shift from agriculture sector to industries and services, the decreasing share of agriculture sector does not affect the importance of this sector in the State economy as the growth in primary sector of the State economy is still determined by the trend in agriculture and horticulture production as it is one of the major contributors to the total domestic product i.e. 13 per cent of state Gross Domestic Product (GDP) and has overall impact on other sectors through employment, trade and transportation, etc. At present agriculture and its allied activities provide employment to 70 per cent of its working population. Out of 55.67 lakh hectare area of Himachal, 9.44 lakh hectare is total operational holding operated by 9.97 lakh farmers (Government of Himachal Pradesh, 2021).

2.2.5 Climate

The climate of Himachal Pradesh, depending on the altitude, varies from semi-tropical to the semi-arctic. It varies from hot and sub-humid tropical (450-900 m) in thc southern low tracts, warm and temperate (900-1800 m), cool and temperate (1900-2400 m) and cold alpine and glacial (2400-4800 m) in the northern and eastern high mountain ranges. The year may be divided into four seasons – the winter season (January to February), the pre-monsoon or hot weather season (March to May), the south-west monsoon (June to September), and the post monsoon season (October to December) as per the IMD (2010).

According to Economic Survey of 2020-21 by Government of Himachal Pradesh (2021), the state has been divided into 4 agro-climatic zones:

- Sub-tropical, sub-mountain and low hills.
- Sub-temperate, sub-humid mid-hills.
- Wet temperate high hills (Kotkhai).
- Dry temperate high hills and cold deserts.

2.3 Shimla

Shimla district of Himachal Pradesh comprises of the 19 small former Shimla Hill States which were merged together after independence. Other than these states, over the years the hilly areas of erstwhile Patiala State and areas of Kotkhai, Kotgarh, Shimla town and Jutogh Cantonment of the provincial government were also included in the district at different stages and came into existence in its present form on 1st September, 1972 (Government of India, 2011).

Shimla district derives its name from the city Shimla which is also the capital of Himachal Pradesh. It is bounded by Kinnaur district in the North-east, by Kullu and Mandi District in the North-west, Solan and Sirmour districts in the South-west and by the State of Uttarakhand in the South-east.

2.3.1 Topography

Except few small valleys, the district is entirely mountainous. The elevation of the district ranges from 300 to 6000 meters. The terrain is mostly comprising of rough, steep valley with high peaks and has thick forest. The major part of the district falls under Zone IV (High Risk Zone) as per the Earthquake hazard zoning of India (District Disaster Management Authority Shimla, 2012).

2.3.2 Administrative units

The total area of the present district is 5,131 sq. kms. There are 10 major towns in the district that are divided on the basis of administrating body, i.e., Municipal Corporation, Municipal Committees, Notified Area Committees and Cantonment Board.

- Municipal Corporation—Shimla
- Municipal Committees—Rampur and Theog
- Notified Area Committees—Choupal, Jubbal, Kotkhai, Narkanda, Rohru, Sunni.
- Cantonment Board—Jutogh

The district comprises of 7 sub-divisions, 12 tehsils, and 6 sub-tehsils. Figure 2.3 is a map that shows District Headquarters and Major Towns in Shimla District.

2.3.3 Demography

Shimla district has a population of 8,14,010 persons with 4,25,039 males and 3,88,971 females; the rural population is 612,659 and urban population is 2,01,351. The district has a population density of 159 inhabitants per square kilometer. As per Government of India Census (2011), Shimla district stands in 10th position in sex-ratio, i.e., 915 females for every 1000 males (rural sex ratio is 949 per 1000 males and urban sex ratio is 818 per 1000 males). Its literacy rate is 83.64 per cent with 89.59 per cent males and 77.13 per cent females and stands 6th in the position in literacy rate (Government of India, 2011).

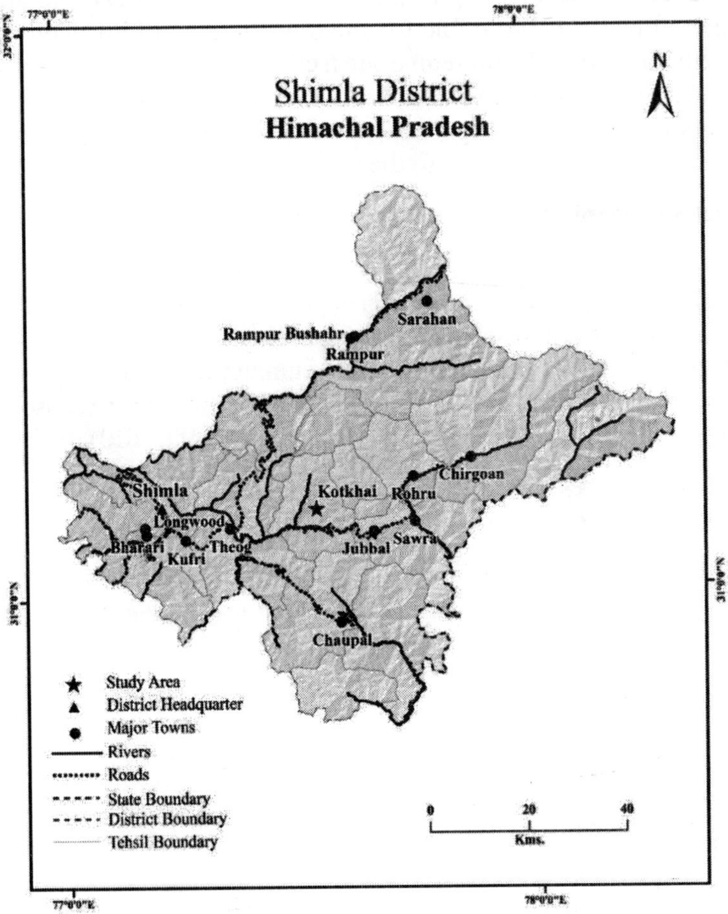

Source: Customised Map.

Figure 2.3: District map of Shimla.

2.3.4 Economy

Economy of the district is dependent to a large extent on horticulture and tourism. It thrives on the revenue from travel and tourism industry and also on the agriculture and horticulture sector. The industries are not well developed in and around Shimla due to its topography, unavailability of proper infrastructure, and high cost of transportation. Tourism provides direct and indirect employment opportunities. The geography and climatic conditions are favourable for the development of horticulture activities especially for temperate fruits. Apple is the main export item for the district and contributes to more than half of the apple production of state. Apple export is one of the major sources of income and contributors to the growth of economy of the state.

2.3.5 Climate

Due to mountainous topography with high ridges, peaks and deep valleys, the climate of the district depends much on elevation. The climate of the district constitutes of fairly cold weather season from December to March, pleasant summer from April to June, brief monsoon from July to mid-September, and a post-monsoon season from mid-September to November (IMD, 2010).

2.4 Kotkhai tehsil

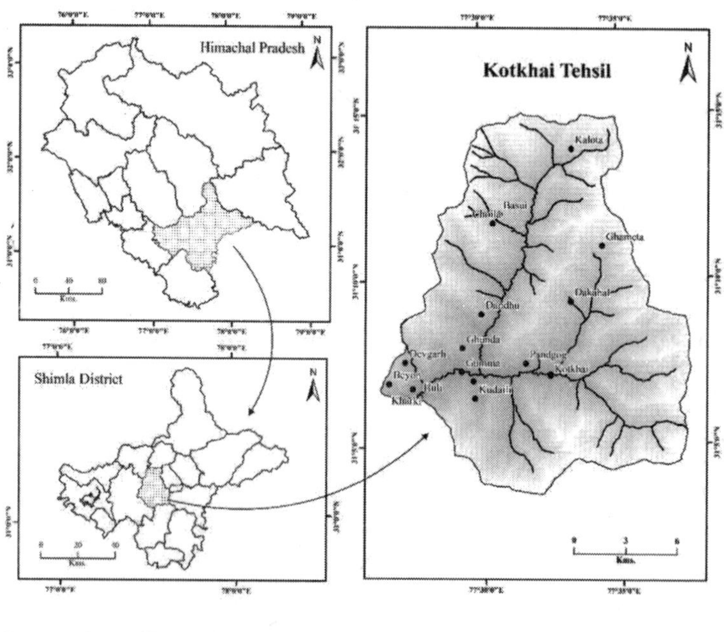

Source: Customised Map.

Figure 2.4: The location of Kotkhai in Shimla (Himachal Pradesh).

Kotkhai is located in District Shimla of Himachal Pradesh (Figure 2.4). It is one of the largest apple-growing valleys in Himachal Pradesh and also a significant contributor to the per capita income of the state. The name Kotkhai is based on location of the palace of its *Rana* (king) and is derived from Hindi and Urdu language, where the name "Kot" means Kings Palace and "Khai" means depth, therefore "Kotkhai" means "Kings palace situated in depth". The whole tehsil is known for quality apple production, its fairs and traditional values. According to the information from Government of India Census (2011), the sub-district code of Kotkhai block is 00189. Total area of Kotkhai is 244 km^2 including 243.24 km^2 rural area and 0.40 km^2 urban area. There are about 250 villages in Kotkhai block, and 24 gram panchayats and one town.

2.4.1 History

Kotkhai was one of the Shimla hill States. It was an erstwhile fiefdom, therefore known as Kotkhai *thakurai*. Kotkhai was said to be founded in 12th AD by the scion of a nearby state Kumarsain, named Ahimal Singh. For large period of time, it remained tributary to Kumarsain state. Owing to its small size, it always remains threatened by the neighbouring larger States such as Bushahr and Kullu. During the Gurkha invasion in 1806, it became their tributaries. Soon after Anglo-Gurkha war in 1815, it was freed from Gurkha rule. In 1826, the Britishers annexed Kotkhai and made it part of British rule in India.

2.4.2 Demography

According to Census (2011), Kotkhai has a population of 39,205 peoples and out of the total population, 38,015 rural population and 1190 urban, i.e., 3 per cent people live in urban areas while 97 per cent lives in the rural areas. The average literacy rate in urban areas is 92 per cent, while in the rural areas it is 82.5 per cent. Also, the sex ratio of urban areas in Kotkhai Tehsil is 722, while in rural areas it is 959. It has lowest sex ratio among towns of Shimla district.

The population of children of age 0-6 years in Kotkhai Tehsil is 4014, which is 10 per cent of the total population. There are 2100 male children and 1914 female children between the age

0-6 years. Thus, as per the Census (2011), the child sex ratio of Kotkhai Tehsil is 911, which is less than average sex ratio (951) of Kotkhai Tehsil. The total literacy rate of Kotkhai Tehsil is 82.77 per cent. The male literacy rate is 80.56 per cent and the female literacy rate is 67.72 per cent in Kotkhai Tehsil. Table 2.2 presents the statistical details of Kotkhai tehsil as per the Census, 2011.

Table 2.2: Statistical detail of Tehsil Kotkhai (Census, 2011)

Particulars	Total	Male	Female
Total population	39,205	20,095	19,110
Children (0-6)	4,014	2,100	1,914
Scheduled Tribe	228	120	108
Scheduled Caste	8,536	4,291	4,245
Literacy Rate	82.77%	80.56%	67.72%

Source: https://www.censusindia.co.in/subdistrict/kotkhai-tehsil-shimla-himachal-pradesh-189

2.4.3 Economy

Kotkhai is known as the apple of Himachal's eyes as it is the major apple producing Tehsil in Himachal, which the major source of the economy. The Baghi-Ratnari valley is the largest apple producing belt of this region. The apple cultivation was first started in Kiari village in mid-1930s in Kotkhai tehsil. Today, the turnover of apple crop in this region is close to thousand crores annually. The success of apple cultivation has brought substantial lifestyle changes among the locals. Kotkhai has one of the highest per capita income in the state owing to the apple cultivation.

Apple cultivation has paved way for other allied activities, which has become major source of income to locals. For example, bee keeping has emerged as a source of livelihood with apple cultivation. An Ag-mark laboratory where honey is processed, packed under Ag-mark label, have been established in Kotkhai. A cold storage of 640MT capacity has been opened in Gumma to store apples, which has provided employment to the local youth. The tehsil has one of the highest numbers of agricultural labourers in district as many people come to the region during apple season in search of livelihood.

Other than apple, other temperate stone fruits are also grown in the area which adds to the economy of the tehsil. These are pears, apricots, plum, peaches and cherry. Vegetables like potato, peas, cauliflower are also grown in lower belts of the tehsil for commercial purpose.

As per Government of India (2011), 22,387 people out of the total population were engaged in some work activities in Kotkhai Tehsil. Out of these 15,474 are cultivators.

2.4.4 Religion

People of Kotkhai are mostly Hindu, however, people of other religions are also present (Table 2.3). About 96.81 per cent of the population of Kotkhai tehsil is Hindu, 2.67 per cent Muslim, 0.09 per cent Christian, 0.02 per cent Sikh, 0.08 per cent Buddhist, 0.21 per cent of other religion, and 0.12 per cent to no specified religion.

The main god/goddess who are known across Kotkhai are four brothers and a sister: *Devatā Baindarā* (the eldest brother), *Devatā Chāmī*, *Devatā Padārā*, *Devatā Ḍīma*, and a sister *Devī Nāndana*. These are the local deities that are equally revered as the Hindu Gods. The people are really devotional towards *Devatā* and every village had its own *Kula Devatā* (clan deity) and *Grāma Devatā* (village deity).

Table 2.3: Religion-wise Population of Tehsil Kotkhai

Religion	Total	Per cent
Hindu	37,954	96.81
Muslim	1,048	2.67
Christian	34	0.09
Sikh	6	0.02
Buddhist	32	0.08
Other Religion	84	0.21
No Religion Specified	47	0.12

Source: https://www.censusindia.co.in/subdistrict/kotkhai-tehsil-shimla-himachal-pradesh-189

2.4.5 Rituals, customs and festivals

Due to the firm belief in *Devatā,* there is a strong holding of *Deva Nītī* in the Kotkhai area and there is a versatility with rich cultural diversity in language, traditions and local deities. The people host various cultural and traditional functions year after year as a devotion to their local deities. Some of these functions are *Khīna, Tikkara, Biśū kā Melā, Sājā, Jātara,* etc.

- *Khīna:* It is celebrated on special occasion in order to worship and thank the deities. It can be organised at family level, or village or community level. The deities are invited to visit the place of the host and are worshipped. Rituals such as animal sacrifice, *hawan,* etc. are performed in order to please the God.
- *Tikkara:* It is a huge festival of Kiari panchayat and is celebrated after four years to express gratitude towards their local deities, *Devatā Baindra, Lankra Veer* and *Mahamaai.* Animal sacrifice was performed prior to ban on the custom of sacrificing animals. The people of the surrounding areas also join in this festival. *Nāṭī* (traditional dance form) and folk songs are performed to celebrate the festival.
- *Biśū kā Melā* (*Biśū* Fair): This is a social and cultural fair, celebrated on the 9th day of April and is continued for the next two days. The fair is held in the honour of *Mahādev Devatā. Nāṭī* (traditional dance form) and folk songs are performed. The devotees pay their homage to the *Devatā.*
- *Sājā: Sājā* (Sankranti) is an auspicious festival that is celebrated four times in a year in the months of January, April, July and August. Various rituals are performed on *Sājā* so that their deity protects them from any inauspicious event, and aids in prosperity of the villagers.
- Mahasu *Jātara* (*Jātara*): This fair is celebrated near Mahasu village on the third Tuesday in the month of May for two days. It is a very old traditional fair and is held in front of the Durga Devi temple and large number of people gather there from the surrounding areas.

One of the common factors in all these rituals and festivals is that the deity too is involved and participate by dancing along with the people if happy.

2.4.6 Language

There are different dialects of *pahāḍī* spoken in different areas of the tehsil, which resemble Kiuthali (Singh, 1961), while Hindi is the lingua franca of the region.

2.4.7 Dress and ornament

The traditional attire of people of Kotkhai tehsil for women comprises of *ḍhātū* (cloth worn on head), *sadrī* (a jacket) and salwar kameez. The men's attire comprises of a jacket, kurta pajama and a *buśaharī topi* (Himachali cap). As the weather is mostly pleasant to cold, woolen clothes are preferred for most part of the year. People of Kotkhai are simple and do not wear ornaments on daily basis. However, most of the men were observed wearing a silver *kaṛā* (wristband) and women wore *kanabāle* (earrings) which were characteristic jewellery articles of the people of field area.

2.4.8 Family structure

The family structure is patrilineal in nature. Traditionally, families were joint in structure, however over the time families have split into separate households. However, nuclear families are rare. The traditional inheritance pattern is in favour of male members. However, over the time, even female members inherit property from their family.

2.5 Field area

The study area for the present study comprises of four villages of Kotkhai tehsil located at different altitudes (Figure 2.5). These villages are Bhawana (1825 m from sea level), Kiari (2224 m from sea level), Up Muhal Kufar Bag (Kufar Bag as known by the locals) (2362 m from sea level) and Up Muhal Jashla (Jashla as known by the locals) (2438 m from sea level).

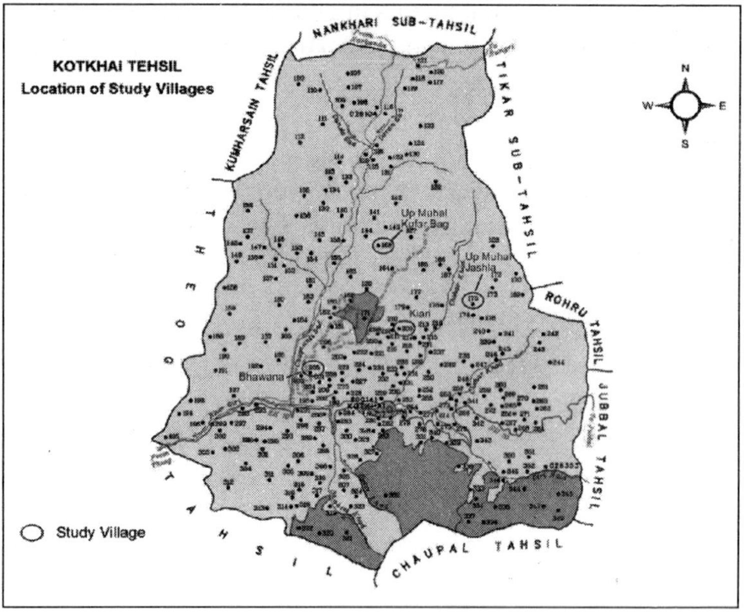

Source: Customized map.

Figure 2.5: Location map of study villages.

2.5.1 Bhawana

According to Census (2011) information, the location code or village code of Bhawana village is 026205. Bhawana village is located in Kotkhai Tehsil of Shimla district in Himachal Pradesh, India. It is situated around 2 km from its sub-district headquarter, Kotkhai, and 56 km away from Shimla city. Bhawana comes under Purag gram panchayat.

The total geographical area of village is 35.47 hectares. According to the Census data (2011), Bhawana has a total population of 199 people, of which 89 are males and 110 are females. There are about 34 houses in Bhawana village.

Bhawana village has lower literacy rate compared to Himachal Pradesh. In 2011, literacy rate of Bhawana village was 82.58 per cent as compared to 82.80 per cent of Himachal Pradesh. In Bhawana male literacy stands at 89.02 per cent, while female literacy rate was 77.08 per cent. Table 2.4 gives the details about the number of houses, population, literacy rate and workers in the village.

Table 2.4: Demographic data of village Bhawana

Particulars		Total	Male	Female
Total No. of Houses		34	-	-
Population	Total	199	89	110
	Child (0-6)	21	7	14
	Scheduled Caste	97	43	54
	Scheduled Tribe	0	0	0
Literacy		82.58%	89.02%	77.08%
Workers	Total	98	46	52
	Main Worker	38	-	-
	Marginal Worker	60	17	43

Source: https://www.census2011.co.in/data/village/26205-bhawana-himachal-pradesh.html

2.5.2 Kiari

Kiari village is located in Kotkhai Tehsil of Shimla district in Himachal Pradesh, India and has a location or village code 026209. It is situated 13 km from sub-district headquarter, Kotkhai, and 70 km away from Shimla city. It comes under the gram panchayat Kiari.

The total geographical area of Kiari is 43 hectares. As per the Census (2011), the village has a total population of 249 people out of which 119 are males and 130 are females. There are about 66 houses in Kiari village.

Kiari village has higher literacy rate as compared to Himachal Pradesh. The literacy rate of Kiari village was 87.67 per cent as compared to 82.80 per cent of Himachal Pradesh. In Kiari, male literacy stands at 94.12 per cent, while female literacy rate was 82.05 per cent (Census, 2011).

The details about total number of houses, population, literacy level and workers in the village have been provided in the Table 2.5.

Table 2.5: Demographic data of village Kiari

Particulars		Total	Male	Female
Total No. of Houses		66	-	-
Population	Total	249	119	130
	Child (0-6)	30	17	13
	Scheduled Caste	26	12	14
	Scheduled Tribe	0	0	0
Literacy		87.67%	94.12%	82.05%
Workers	Total	90	69	21
	Main Workers	80	-	-
	Marginal Worker	10	9	1

Source: https://www.census2011.co.in/data/village/26209-kiari-himachal-pradesh.html

2.5.3 Kufar Bag

The location code or village code of Kufar Bag village is 026168s per Census (2011). It is located in Kotkhai Tehsil of Shimla district in Himachal Pradesh, India. It is situated 22 km from the sub-district headquarter in Kotkhai and 74.4 km from Shimla city. It comes under Parala Badrooni gram panchayat.

The village covers approximately 98 hectares of geographical area. According to Census (2011), Kufar Bag has a total population of 103 people, out of which 50 are males and 53 are females. There are about 28 houses in the village.

Kufar Bag village has lower literacy rate as compared to Himachal Pradesh. As per Census (2011), literacy rate of Kufar Bag village was 79.57 per cent as compared to 82.80 per cent of Himachal Pradesh. The male literacy of the village stands at 86.67 per cent, while the female literacy rate was 72.92 per cent. Table 2.6 depicts the number of houses, population, literacy rate, and worker population in the village.

Table 2.6: Demographic data of village Kufar Bag

Particulars		Total	Male	Female
Total No. of Houses		28	-	-
Population	Total	103	50	53
	Child (0-6)	10	5	5
	Scheduled Caste	6	3	3
	Scheduled Tribe	0	0	0
Literacy		79.57%	86.67%	72.92%
Workers	Total	39	30	9
	Main Worker	39	-	-
	Marginal Worker	0	0	0

Source: https://www.census2011.co.in/data/village/26168-up-muhal-kufar-bag-himachal-pradesh.html

2.5.4 Jashla

Jashla is located in Kotkhai Tehsil of Shimla district in Himachal Pradesh, India with a location or village code 026175. It is situated 75.7 km from the city of Shimla and 18.6 km away from the sub-district headquarter in Kotkhai. It comes under Devri Khaneti gram panchayat.

Table 2.7: Demographic data of village Jashla

Particulars		Total	Male	Female
Total No. of Houses		127	-	-
Population	Total	542	280	262
	Child (0-6)	54	25	29
	Scheduled Caste	122	62	60
	Scheduled Tribe	0	0	0
Literacy		80.33%	89.41%	70.39%
Workers	Total	406	212	194
	Main Worker	188	-	-
	Marginal Worker	218	33	185

Source: https://www.census2011.co.in/data/village/26175-up-muhal-jashla-himachal-pradesh.html

The total geographical area of village is 167 hectares. Jashla has a total population of 542 peoples, of which 280 are males and 262 are females. There are about 127 houses in Jashla village (Census, 2011).

The literacy rate of village is low in comparison to overall literacy rate of the state. In 2011, literacy rate of Jashla village was 80.33 per cent as compared to 82.80 per cent of Himachal Pradesh. In Jashla male literacy stands at 89.41 per cent, while female literacy rate was 70.39 per cent. The demographic data about the total number of houses, population, literacy and workers have been demonstrated by Table 2.7.

3

Climate Change and People's Perception

This chapter comprises of analysis of the available climatic data for Kotkhai tehsil from the secondary source, i.e., Indian Meteorological Department (IMD) as well as people's perception about climate change and its impact on their day to day lives. This section comprises of the first two objectives of the present study, i.e., to analyse climatic data (precipitation and temperature) of study area, and to study the change in climatic conditions and its impact on day to day activities in the field area.

3.1 Analysis of climatic data of the study area

Available data regarding temperature, rainfall and snowfall, were collected from the Meteorological Centre (IMD), Shimla, and analysed to find the climatic trends for past few decades.

3.1.1 Temperature

The analysis of the data provided by Meteorological Centre, Bibra House, Cliffend Estate, Shimla (Himachal Pradesh), for past fifty years (1971-2020) is portrayed in Figure 3.1 and 3.2. The temperature has been increasing constantly for the past five decades. Overall, the decades 2011-2020 has been the warmest in past five decades. The mean maximum temperature for the decade 1971-1980 was 17.43°C which increased in every decade and measured 20.1°C in 2011-2020. A remarkable increase of around 2.67°C has been observed in the mean maximum temperature for the past 50 years, as can be seen in Figure 3.1.

Similarly, the mean minimum temperature has risen from 9.41°C in 1971-1980 to 13.5°C in 2011-2020. A sharp increase in mean minimum temperature has been observed in the last decade, 2011-2020 (13.5°C), from the decade 2001-2010 (11.56°C), as displayed in Figure 3.2.

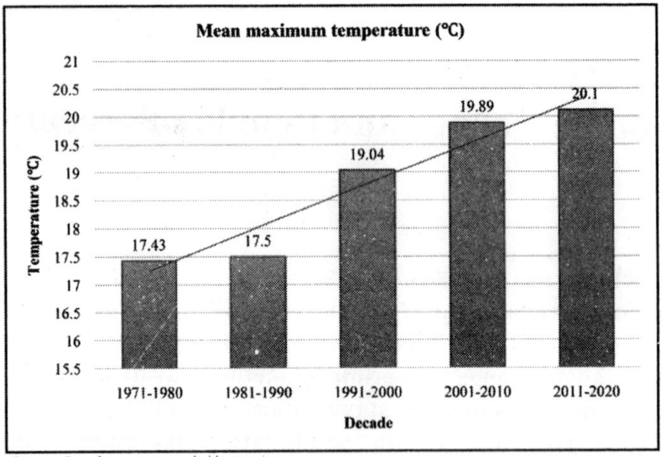

Source: Based on data provided by IMD.

Figure 3.1: Mean maximum temperature (in °C) of five decades (1971-2020).

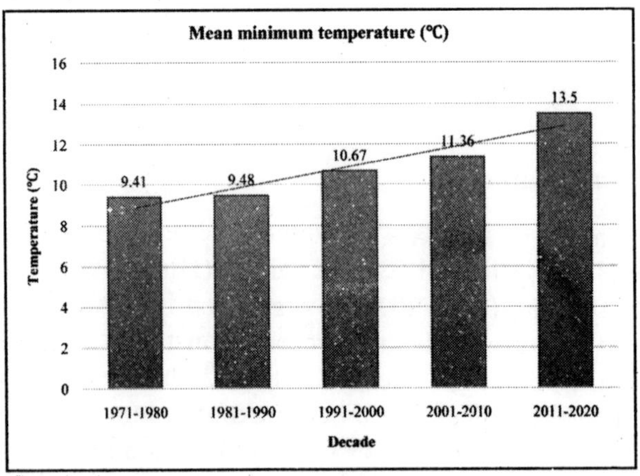

Source: Based on data provided by IMD.

Figure 3.2: Mean minimum temperature (in °C) of five decades (1971-2020).

The general findings of the study regarding climate trend for temperature, i.e., an increasing trend, corroborates with the findings of Stafford *et al.* (2000) who recorded temperature from 1949 to 1998 to conduct a 50-year trend analysis of the climate variables. They found that the annual and seasonal mean temperature showed an increasing trend throughout Alaska. A study by Hayelom *et al.* (2017) which was based on the analysis of temperature from 1985-2010, found that temperature presented a general increasing trend in southern Tigray regional state (Ethiopia) which is in accordance to the upward trend in temperature from 1971-2020 as noted in the present study. The results of the present study also concords with the Intergovernmental Panel on Climate Change (IPCC, 2021) according to which each of the last four decades has been consecutively warmer than any decade that preceded it since 1850.

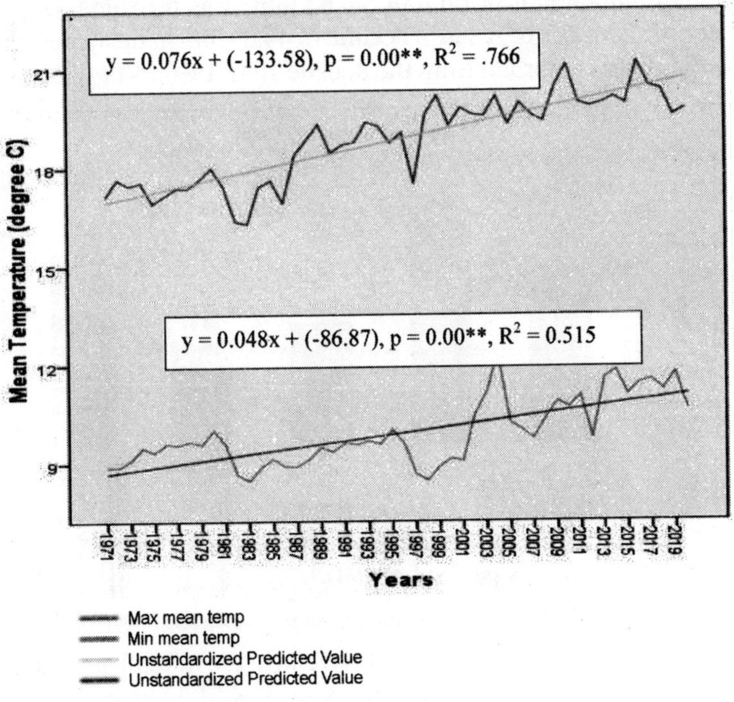

Figure 3.3: Linear regression trendline for maximum and minimum temperature (in °C).

The linear regression trendline (Figure 3.3) shows a significant increase in maximum and minimum temperature from 1971 to 2020. The maximum temperature is increasing at a rate of 0.076 °C/year [y = 0.076x + (-133.58), p = 0.00**, R2 = 0.766], and minimum temperature is increasing at a rate of 0.048 °C/year [y = 0.048x + (-86.87), p = 0.00**, R2 = 0.515]. The highest mean maximum temperature was recorded for the year 2016 followed by 2010, and highest mean minimum temperature was recorded for the year 2004.

3.1.2 Rainfall

The fifty-year rainfall data for Kotkhai recorded an irregular pattern for past five decades, i.e., 1971-2020 and has been presented in Figure 3.4. As is evident from Figure 3.4, the decade 1981-1990 observed the maximum mean rainfall (96.5 mm), while minimum amount of rainfall (44.62 mm) was recorded for the decade 2001-2010. A sharp decline of 44.09 mm in mean decadal rainfall was recorded from the decade 1981-1990 to 1991-2000.

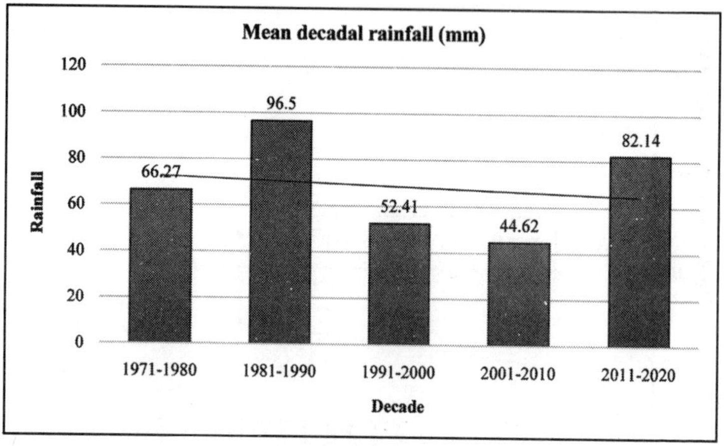

Source: Based on data provided by IMD.

Figure 3.4: Mean rainfall (in mm) for the five decades (1971-2020).

There has been a non-uniform trend in the rainfall pattern for pre-monsoon (months of March, April and May) as depicted in Figure 3.5. Interestingly, there has been a significant decrease in rainfall from 137.16 mm in 1981-1990 to 33.13 mm in 1991-2000.

The last decade, i.e., 2011-2020, received second highest mean rainfall of 57.99 mm in past fifty years, which is remarkably less than the mean rainfall of 1981-1990.

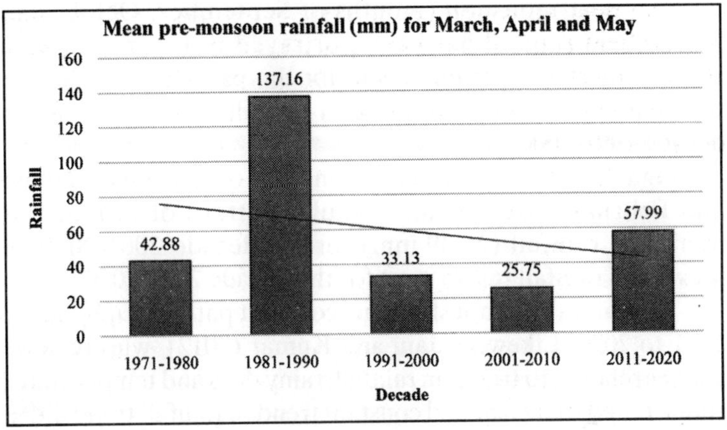

Source: Based on data provided by IMD.

Figure 3.5: Mean pre-monsoon rainfall (in mm) for the five decades (1971-2020).

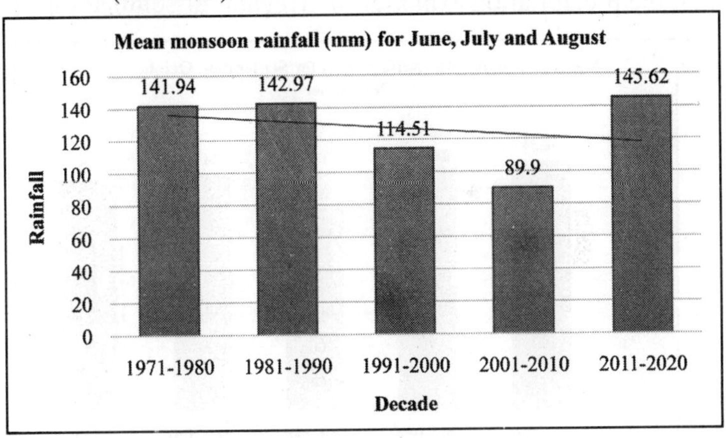

Source: Based on data provided by IMD.

Figure 3.6: Mean monsoon rainfall (in mm) for the five decades (1971-2020).

Figure 3.6 demonstrates rainfall in monsoon (months of June, July and August), where a gradual decline in rainfall has been observed from the decade 1981-1990 to the decade 2001-2010, thereafter a significant increase in rainfall for the decade 2011-2020 (145.62 mm) is noticed, which is also the highest recorded mean monsoon rainfall in the past fifty years.

The post-monsoon (months of September, October and November) rainfall has been portrayed in Figure 3.7 which shows almost comparable result for the past five decades with the minimum mean post-monsoon rainfall of 31.12 mm for the decade 1991-2000.

Figure 3.8 for winter rainfall (months of December, January and February) depicts an irregular pattern of rainfall with minimum rainfall (26.39 mm) for the decade 2001-2010 and maximum rainfall (69.75 mm) for the decade 2011-2020.

The rainfall did not show any constant pattern from the year 1971 to 2020. Likewise, Jain and Kumar (2012), who reviewed studies relating to trends in rainfall, rainy days and temperature in India, noted that clear and constant trend of rainfall trend did not emerge. Similar observations were made by Hayelom *et al.* (2017) who analysed precipitation of thirty years, from the year 1981-2010, in southern Tigray regional state (Ethiopia), and observed that the precipitation exhibited an irregular dissemination.

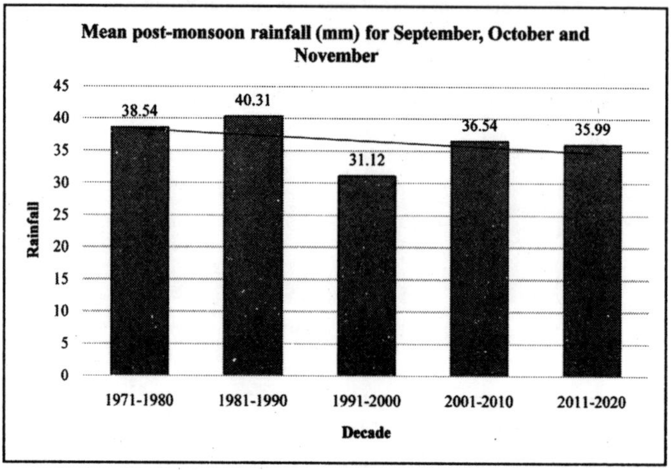

Source: Based on data provided by IMD.

Figure 3.7: Mean post-monsoon rainfall (in mm) for the five decades (1971-2020).

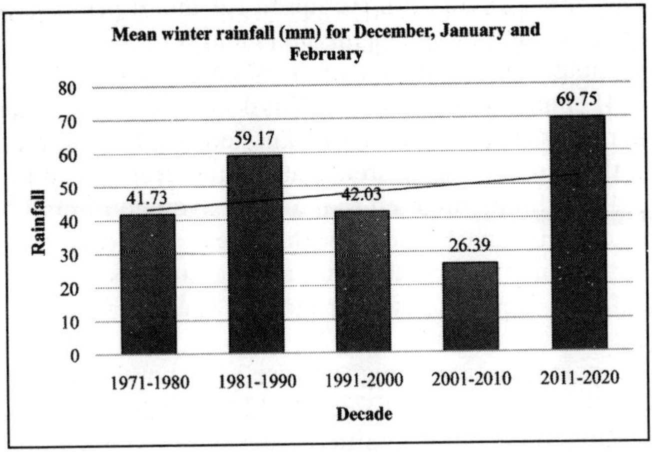

Source: Based on data provided by IMD.

Figure 3.8: Mean winter rainfall (in mm) for the five decades (1971-2020).

The linear regression trendline for mean rainfall (Figure 3.9) for fifty years shows a decreasing rate of -0.371mm/year [y = -0.371x + 808.14, p = 0.19, whereas $R^2 = 0.515$]. The curve reveals inconsistent declining rainfall with maximum mean rainfall in the year 1990, and minimum in the Year 2013.

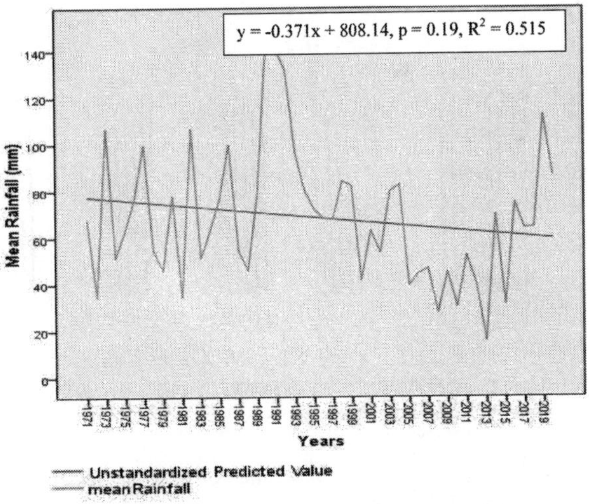

Figure 3.9: Linear regression trendline for annual mean total rainfall (in mm).

The linear regression trendline for mean pre-monsoon (Figure 3.10), monsoon (Figure 3.11) and winter (Figure 3.13) rainfall noted a declining rate of -0.487 mm/year, -1.26 mm/year and -0.412 mm/year respectively. This decreasing trend is similar to the results that were found in a meteorological monograph about the climate change trends at a state level by Rathore *et al.* (2013), who highlighted that annual average rainfall have declined in Himachal Pradesh with highest significant decrease in monsoon months rainfall in July among all states in India, and a significant decrease in winter months rainfall in January for Himachal Pradesh.

The post-monsoon rainfall (Figure 3.12) shows an increasing rate of 0.327 mm/year [$y = 0.327x + (-612.61), p = 0.31, R^2 = 0.022$]. The results depicted a significant decline in mean monsoon rainfall [$y = -1.26x + 2662.4$, $p = 0.02^*$, $R^2 = 0.102$]. It can be inferred from the results that there has been a seasonal shift in rainfall from monsoon towards post-monsoon.

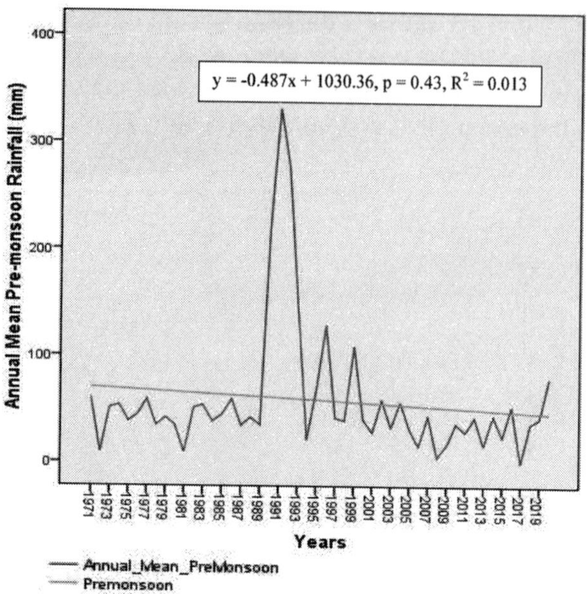

Figure 3.10: Linear regression trendline for annual mean pre-monsoon rainfall (in mm).

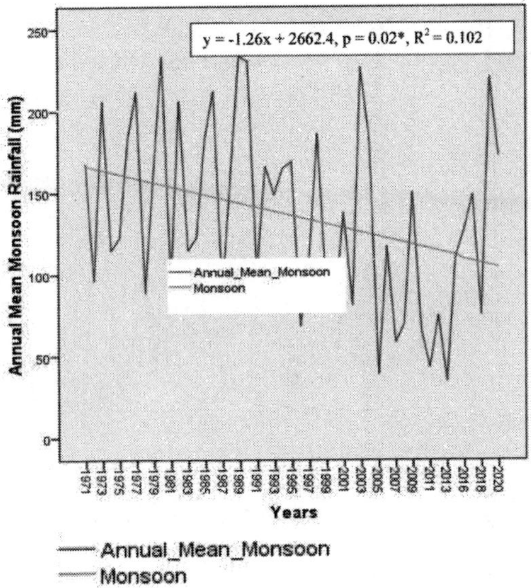

Figure 3.11: Linear regression trendline for annual mean monsoon rainfall (in mm).

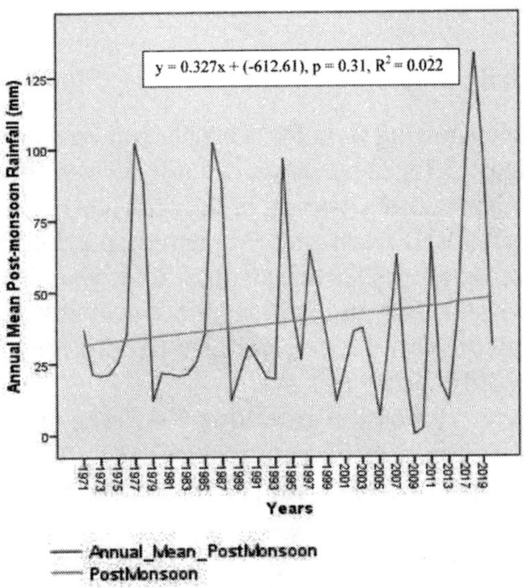

Figure 3.12: Linear regression trendline for annual mean post-monsoon rainfall (in mm).

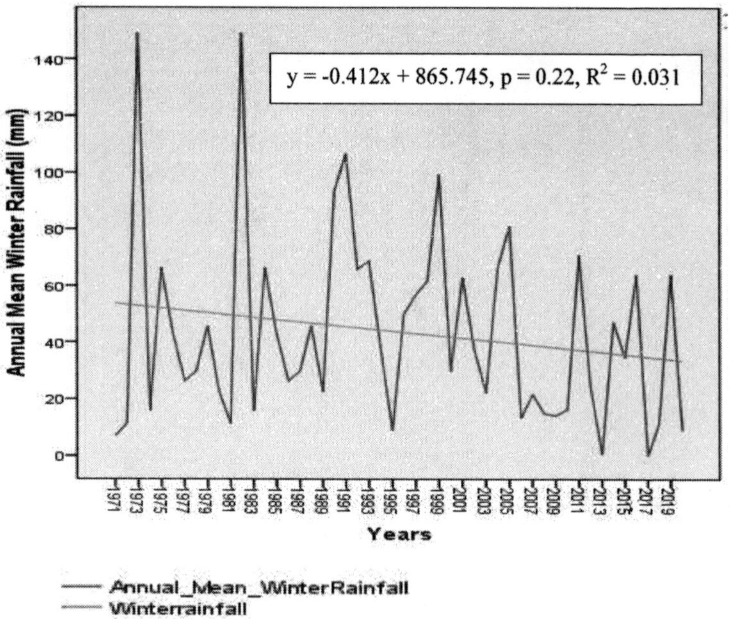

Figure 3.13: Linear regression trendline for annual mean winter rainfall (in mm).

3.1.3 Snowfall

The available snowfall data for Kotkhai was analysed in Figure 3.14 and Figure 3.15. The pattern of snowfall was observed to be irregular with maximum amount of snowfall (cm) in the year 2020 since the year 2010 (Figure 3.14). Maximum snowfall (34 cm), was recorded for the year 2020 and minimum for year 2010 (0.5 cm). Data for the year 2015 and 2018 were not available for snowfall. The snowfall displayed a non-uniform pattern for the past one decade, i.e., 2010-2020.

The linear regression trendline for mean snowfall in cm (Figure 3.15) for past eleven years (2010-2020) reveals an increasing rate of 0.391 cm/year [$y = 0.391x + (-783.81)$, $p = 0.288$, $R^2 = 0.124$].

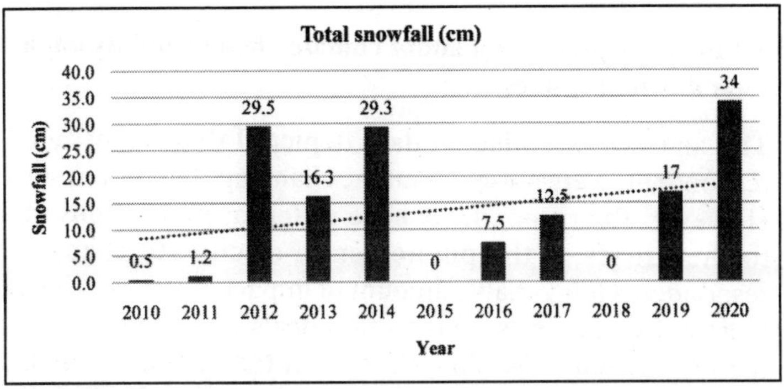

Source: Based on data provided by IMD.

Figure 3.14: Annual mean snowfall (in cm) for the years 2010-2020.

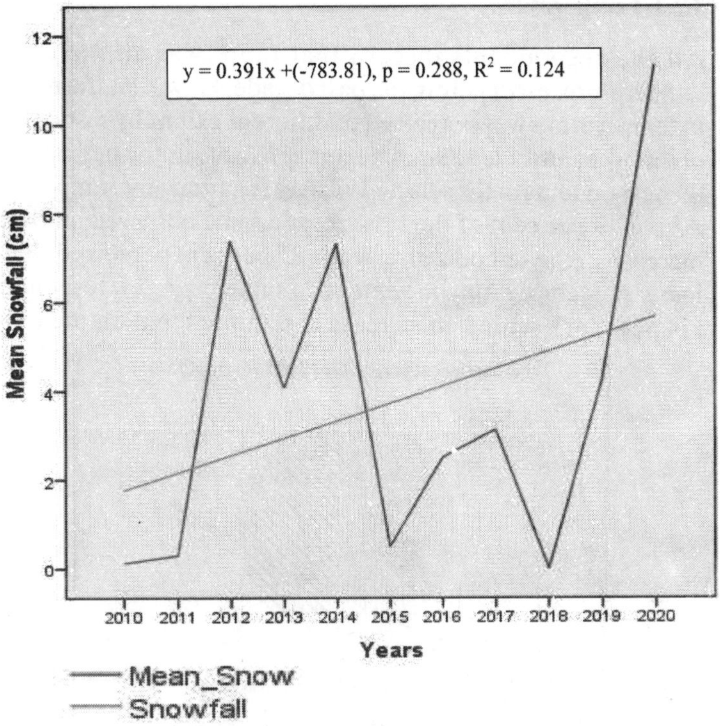

Figure 3.15: Linear regression trendline for annual mean snowfall (in cm) from 2010 to 2020.

3.2 People's perception about climate change and its impact on their day to day lives

The analysis of collected data depicted that majority of the respondents were aware about the changing climatic parameters. However, the respondents have different perceptions related to the causes of the phenomenon of climate change. The respondents felt variable amount of impact of climate change on their day to day lives. This section deals with the perception of the respondents regarding changes in temperature rainfall and snowfall, as well as their perception about the cause of climate change, factors affecting their perception, and impact of climate change on their day to day lives.

3.2.1 Temperature

All the respondents discern that there was an increase in temperature, especially in the past decade. However, the increase in temperature was perceived to different extent by respondents of different altitude villages. Figure 3.16 reveals the perception of the respondents of the selected villages regarding the temperature. About 49 per cent of the tctal respondents observed an overall increase in the temperature, while 37 per cent perceived slightly increased temperature in winter and summer. A few respondents (14 per cent), sensed an increase in summer temperature only.

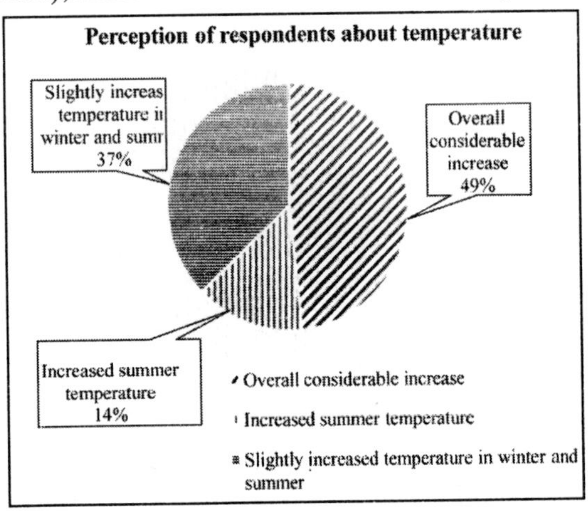

Figure 3.16: Respondents' perception regarding temperature.

The lower and lower-middle altitude village respondents observed that there was a significant increase in temperature. A respondent (82-year-old, male, from the lower middle altitude village, Kiari), was of the opinion that the temperature has increased drastically over the years. He observed that summers are harsher now and winters are not as cold as they used to be. According to a respondent (51-year-old, male, from low altitude village, Bhawana), the temperature had at least increased by 2-3 degree. While, the lower and lower-middle altitude villages of Bhawana and Kiari felt harsher temperature conditions, increasing trend was observed by the respondents of upper-middle and high-altitude villages, i.e., Kufar Bag and Jashla respectively, as well. A respondent (75-year-old, female, from the upper-middle altitude village, Kufar Bag) have seen an increase in heat intensity in summer. A respondent (59-year-old, female, from the high-altitude village, Jashla) noted that there have been changes in the temperature during winter as well as summer. Earlier, the winters were bitter and harsh and summers were pleasant. However, now the winters are becoming mild and scorching heat could be felt in summers.

3.2.2 Rainfall

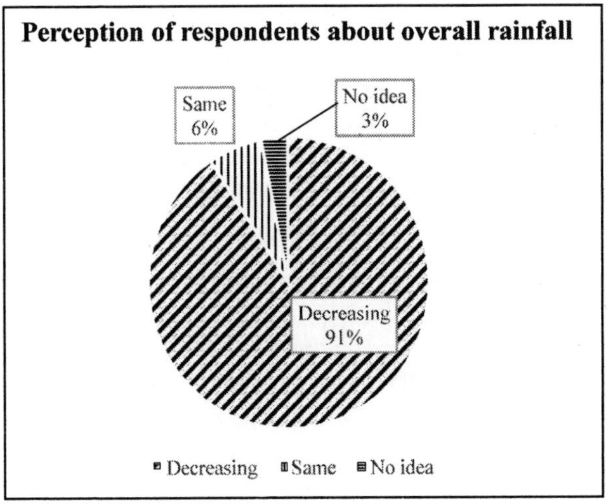

Figure 3.17: Respondents' perception regarding overall rainfall.

Perception of all respondents regarding overall rainfall is portrayed in Figure 3.17. A considerable percentage of the respondents (91%), felt that there was a decline in rainfall over the years. A little proportion of participants (6%) noted that there was no change in rainfall, while about 3 per cent respondents had no idea about the rainfall pattern.

The respondents noticed that though there was an overall decline in rainfall, there were erratic bouts of rainfall, sometimes accompanied by hail. A respondent (in her late 80s, female, from the lower-middle altitude village, Kiari) has observed a decline in rainfall which was almost negligible now. Another respondent (74 years old, male, from the upper-middle altitude village, Kufar Bag) stated:

> *Hamāre zamāne me to bāriśa lagātāra hotī thī. Aba to bahāne ke liye pānī barastā hai.*

During our time it used to rain continuously at same rate, however, now it rains to run off.

Most of the respondents noticed that the rainfall was inconsistent and untimely. It did not rain like it used to during monsoon. Intensity of post-monsoon showers had increased and pre-monsoon showers were delayed. Also, earlier there used to be some rainfall and mild hailstorm as the winters passed and spring arrived in March. But now, the rainfall followed no particular pattern, and random and intense hailstorm had become frequent throughout the year. According to a respondent (51-year-old, male, from the lower altitude village, Bhawana) there was a drastic decline in local rain like *Jeṭhāgari* (summer rain in month of *Jeṭha,* mid-May to mid-June),

> *Pahale garmīyo me bahuta bhārī bāriśa hūa kartī thī, 20-30 minutes ke liye hi, para itnī heavy rain hotī thī vo kī nāle bahā karate the. Jaba hama school jāyā karate the to gharvāle bolate hote the kī Jeṭhāgari hai dhyāna se jānā. Jaba bāriś hotī ūpara kī side jaise Kufar Bag me, to patā hotā thā kī āja to nāle bahenge.*
>
> Earlier there used to be heavy rainfall during summer, for 20-30 minutes, but so heavy that streams used to

> flow. When we used to go to school, the family member used to warn us about *Jeṭhāgari* (summer rain). When it used to start raining at villages above, like Kufar Bag, we knew that today streams will flow.

Thus, a decline in summer rain or pre-monsoon rainfall was observed by the respondents of the field area.

The overall decline in rainfall as observed by the people is similar to the findings of Bhutiyani *et al.* (2007) who stated that the annual precipitation, winter and monsoon precipitation over Shimla for the time period 1866-2006 have followed a decreasing trend. The overall decline in rainfall over the past decades confer with the study of HP-SAPCC (2012), which presents that district of Shimla and Sirmour had a decreasing trend of rainfall for past 25 years and increasing trend for district Kinnaur, Chamba and Lahul and Spiti.

However, the general decline in rainfall over the years observed in the present study is in divergence with the findings of IPCC (2021) which recorded an increase in the precipitation since 1950s. The study by IPCC (2021) is a global study and gives the average for the whole world; this may explain the variation from the results of present study.

3.2.3 Snowfall

All the respondents observed a decline in snowfall over the years. One of the respondents (46 years old, male, from the lower altitude village, Bhawana) noted a major decline in snowfall since his childhood. According to him, it used to snow around 1 to 2 ft earlier, while there was no snowfall at all now. A respondent (75-year-old, female, from the upper-middle altitude village, Kufar Bag) recalled:

> *Jaba main āyī thī yahān taba sardiyon me ye kamara taka barpha paṛā karatī thī. Bāhara bhī nahīn jā sakate the. Gāya duhane bhī obare phōṛa se hokara jānā paṛatā thā.*
>
> *When I came here, it used to snow till our waist (a symbol of high snowfall) during winter. We could not even go outside. We had to go to cowshed for milking the cow through phōṛa, a room with a passage that connected the house to the cowshed.*

It used to snow to an extent that they could not even go outside from their house, and this is one of the reasons why in old houses, the cowshed was right below the house and accessible from within the house.

The respondents from Bhawana witnessed years when there was no snowfall at all. The respondents from Kiari observed negligible snowfall for past decade. Other than perceiving a definite decline in snowfall over the time, a change in timing of snowfall was also observed. According to the respondents, while the snowfall started around end of December in the 90s, the timing shifted to end of January by the mid-2000s. A respondent (33-year-old, female, from the upper-middle altitude village, Kufar Bag) recalled:

> *Mujhe yāda hai bachapana me hama chaire lagā rahe hote the barpha me jaba* Christmas *ke āsa pāsa hamārā report card ātā thā.*
>
> I remember how during childhood we used to be sledging in snow when our report card used to arrive around Christmas.

Another respondent (58-year-old, women, from the lower-middle altitude village, Kiari) cited that:

> *Pahale kī bāta alaga thī. Disambara me hī kītanī barpha paṛa jāti thī. Abhī bhī chhabbī Janavarī taka paṛa hi jāti thī. Para kuchha sālon se koi hisāba hi nahīn hai. Kabhī Faravarī me paṛa rahī, kabhī Mārcha me, vo bhī nāma mātra.*
>
> It was different before. It used to snow so much in December. Till some time back, it used to snow around 26th of January. But from past few years there has been no estimation. Sometimes it snows in February, sometimes in March, that too negligible.

Thus, the respondents observed not only a decline in snowfall, but also a shift in timing of snowfall. The snowfall pattern has been becoming erratic according to the respondents. The perception of people regarding decline in snowfall has been supported by study of De *et al.* (2005), who observed that different altitudes in Himachal have shown a decreasing trend of snowfall at Bhang, Solang, Dhundi and Patseo station for recent 20 years of observations.

3.2.4 People's perception about cause of climate change

The people of the field area believed that there were various reasons behind the change in climate. Some believed that it was nature and their deities punishing them for misusing the resources and not keeping up with their culture, while many believed that it was a global phenomenon that was accelerated by various anthropogenic activities like deforestation, burning of fossil fuels, changes in the way the land was used, unsafe garbage disposal and building concrete buildings instead of traditional eco-friendly houses. Figure 3.18 depicts the various factors, like natural and man-made, that led to climate change, as perceived by the respondents.

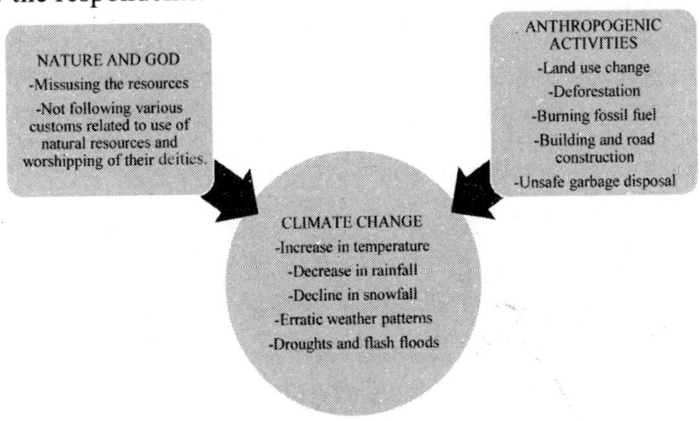

Source: Field data.

Figure 3.18: People's perception regarding factors leading to climate change.

A respondent (in her late 80s, female, from the lower-middle altitude village, Kiari) felt that:

> *Sūkhā paṛanā, barpha nā paṛanā, saba Bhagavāna kā khela hai. Vo dekha rahā hai kaise hamāre bachche mā bāpa kā dhyāna nahīn rakha rahe nā apnī buzurgon kī zamīna kā dhyāna rakha rahe.*

> Dry spell, no snowfall, all these are the play of God. He is seeing how the generations after her are not taking good care of their parents and their ancestral land.

According to a respondent (82-year-old, female, from the lower-middle altitude village, Kiari), there is a change in the climatic conditions and observed that the weather is worse in the recent times. She believes that:

> *Mausama kā badalāva Bhagavāna kā kīyā huā hai. Duniyā aise hi khatama hogī. Kalayuga kā anta ā rahā hai. Achānaka se bhāri barasāta honā, sūkhā paṛanā, bahuta garmī paṛanā, ye saba Bhagavāna sazā de rahe hai logon ko unake kīye kī jo ye ek dūsare aur prakriti ke sātha kara rahe.*
>
> Climate change is an act of God. This is how the world will come to an end. The present time, *Kaliyug* (last stage out of the four stages that the world goes through in Hinduism) is about to come to an end. The sudden excessive rainfall and random drought like conditions, harsh summers, etc., are the signs that the people are being punished for their atrocities against each other and nature.

According to a respondent (male, 80 years old, from the upper-middle altitude village, Kufar Bag):

> *Barpha bahuta kama paṛa rahī, bāriś kabhī bhī ho jātī. Mausama badala rahā kyonki ādamī badala rahā. Seba ke sāth paisā ā gayā hai to lālachi aur ghamanḍī ho gaye hai loga. Loga prakriti kā nājāyaz istemāla karate apane matalab ke liye, Devatā to narāza honge hī. Ye saba mausama kā khela Devatā kā kīya huā hai ādamiyon kī galatiyo kī vajaha se. Ūpara se āja kala ke bachche mānate hī nahīn kuchha.*
>
> There has been a decline in snowfall and rainfall is becoming erratic. The weather is changing because of the change in men's nature/behaviour. With the progress of cash crop economy over the years they

> become greedy and arrogant. People misuse nature for their selfish interests, thus making the deities angry. These changing climatic conditions are a way of their deities to punish people for their mistakes. Also, the children these days do not believe in anything.

The above narratives points at the belief system of the respondents towards the role of deities in changing the climate. Other than the perception that climate change was an act of God, majority of the respondents believed that the anthropogenic activities caused climate change.

A respondent (51-year-old, male, from the lower altitude village, Bhawana) observed that:

> *Ye on-going* climate change *eka global phenomena hai jo ādmiyon dvārā ho rahā. Ye jo chīze* climate change *kara rahī nukasānadāyī to hai par zarūrī bhī hai, jaise saṛake ho gaī, loga jungle kāṭa rahe saṛaka banāne ke liye tākī bākī duniyā se* connect *ho sake, yā lakaṛiyạ̄ jalā rahe sardiyon me garma rahane ke liye. Aba* nuclear family *kā* trend *bhī ā gayā hai usakī vajaha se* resources *jaise* land, *unapar zyādā* pressure *paṛa rahā kyonki sabako kamānā hai.* Developing country *ke hone kī vajaha se hamārā itanā yogadāna nahīn hai jitanā* developed countries *kā hai, para* unplanned development *aura* urbanization *ke chakkara me* concrete *ke jungle banā rahe hai hama.*

> On-going climate change is a global phenomenon caused by the anthropogenic activities which are detrimental but needed, for instance, the construction of roads by cutting down the forests to make the area accessible to outside world and better facilities, or burning of wood to keep themselves warm during long winter. The change in social structure, like nuclear family instead of joint family has put more pressure on the resources as land is being divided and exploited for more benefits. Though the accountability of developing nations in climate change is not as much as of the developed nations, but unplanned development

such as random urbanisation has contributed to it, which is nothing but formation of concrete jungles.

Another respondent (Male, 82 years old, from the lower-middle altitude village, Kiari) have witnessed a drastic change in climatic parameters over the time. According to him:

> *Seba kī bāgavānī ke chalate logo ne binā soche samajhe bahuta jungle kāṭa diye. Ghāsniyon para bhī ab bagīche khaṛe hai jisse zamīn kī ūrjā khatma ho rahī. Moharū aura bāna ke peṛa kāṭa dīye logo ne jungle ke sātha jo āsa pāsa namī bānaāye rakhate the. Iskī vajaha se sūkhā paṛane lagā hai aura kahin nā kahin ina saba chīzon se mausama me badalāva ā rahā hai.*
>
> Apple cultivation has led to unplanned and unprecedented deforestation. Ghasaniyan (the grasslands) have been converted into apple orchards, thus putting land under more pressure. During deforestation many trees like moharū (Quercus dilatata), and bāna(Quercus leucotrichophora) were cut, which used to maintain the moisture in the surroundings. This has led to drying conditions and have led to climate change in some way or the other.

A respondent (65-year-old, male, from the upper-middle altitude village, Kufar Bag) observed that:

> *Āja se dasa sāla pahale* same *pahāṛo para purā sāla barpha hotī thī. Juna Julaī me bhī in* ranges *para barpha dikhatī thī. Aba* global warming *se garamī kī vajaha se saba badala gayā. Ūpara se logo ne peṛa kāṭa kara aura sūkhā kara diyā hai. Kīari ke ūpara* Kalada jungle *pahale* sacred forest *thā. Koī peṛa nahīn kāṭatā thā. Phira logo ne khuda hī khola diyā.* Deforestation *huī,* climate change *huā, aba* traditional *bāoṛiyān usa* area *me sūkha gaī hai.*
>
> Ten years from now same mountains were covered with snow for the entire year. Snow could be seen in June-July on these ranges. Now due to global warming,

> everything has changed. Above that people have dried the area by cutting the trees. Kalada forests above Kiari was a sacred forest before. Nobody would cut the trees. Then people themselves opened it. Deforestation happened, climate change occurred, now traditional *bāoṛiyān* of that area has dried out.

A respondent (54-year-old, female, from the upper-middle altitude village, Kufar Bag) believed that the major cause of climate change is the deforestation and forest fire, which is very prominent in their area. One of the respondents (37-year-old, male, from the high-altitude village, Jashla), viewed climate change as a global phenomenon caused by burning of fossil fuels and carbon emission by industries which has led to decline in amount of snowfall and rainfall and an increase in temperature.

Thus, respondents were of the opinion that anthropogenic activities such as deforestation and burning of fossil fuel were the leading cause of climate change. There were also respondents who had observed changes in climate over the years, but did not know the cause of it. For instance, according to a respondent (75-year-old, female, from the high-altitude village, Jashla),

> *Mausama badala to rahā hai para kīsako patā kyon badala rahā. Prakriti kā khela prakriti hī jāne.*
>
> (Weather is changing but who knows why. Nature only knows its course/game).

3.2.5 Factors affecting the perception about climate change

The factors affecting the perception about climate change can be broadly divided into three categories (Figure 3.19).

- **Modernisation and Globalisation:** Given that in today's time, almost everyone in the field-area were using mobile phones, television and some were also using social media and internet, and climate change being the debatable topic that it has become over the time, people were aware about the term climate change. They had felt the impact at personal level as their livelihood was affected in some way or the other.

- **Education:** Respondents with higher education were of the belief that climate change was due to on-going human activities such as burning of fossil fuel, deforestation for development and personal use. While the respondents with less or no educational qualification blamed changing dynamics between the man and nature for unpredictable climatic conditions. Though not every respondent who was illiterate believed that the climate change was the result of punishment by God. These people were aware about the scientific reasons behind the climate change due to the ongoing debates on television and other social platforms.
- **Traditional and scientific theories:** Some of the respondents believed that climate was in the control of their deities and rituals performed on various occasions like *Sājā* (Sankranti), *Khīna*, and *Biśū kā Melā* kept their deities happy thus averting any incidents such as bad weather conditions, hailstorm, excessive rainfall, or no snowfall. Other respondents did not believe in such traditional beliefs and practices and blamed the activities at local and global level such as burning of fossil fuels and deforestation for climate change.

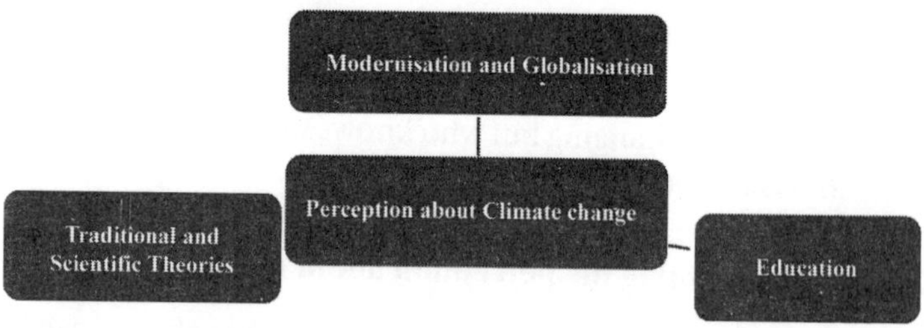

Source: Field data.

Figure 3.19: Factors affecting the perception about climate change.

All the respondents were aware of the changes in various climatic parameters over the time. They had observed an increase in temperature and a decline in rainfall and snowfall over the years. According to them, the events of erratic heavy rainfall and hail were becoming frequent recently. The unseasonal events like no snowfall during winter, very high temperature and random lashes of rainfall made the ongoing global phenomenon of climate change felt strongly by the people. The traditional calendar with expected and actual weather events have been displayed in Table 3.1.

Table 3.1: The traditional calendar with expected and actual weather events

Local name	**Month**	**Expected weather**	**Actual weather**
Māgha	15th January- 14th February	Cold and snowfall	Cold and erratic snowfall
Phālguna	15th February- 14th March	Cold with snowfall and hail	Less cold and unexpected precipitation
Chaitra	15th March- 14th April	Occasional hail and pleasant	Hot and erratic rainfall and hail pattern
Baśaeo	15th April- 14th May	Warm and clear weather	Hot and random spells of rain
Jeṭha	15th May- 14th June	Hot and dry	Hot and erratic rain and hail
Ṣāḍ	15th June- 14th July	Pre-monsoon rainfall	Hot and erratic rainfall
Śāvana	15th July- 14th August	Monsoon rainfall and humid	Hot and humid with bouts of rainfall
Bhādo	15th August- 14th September	Monsoon rainfall and humid	Monsoon rainfall and humid
Ṣauja	15th September- 14th October	Dry air and temperature go down	Random bouts of rain
Kātī	15th October- 14th November	Dry and cold	Dry and cold
Mangaṣara	15th November- 14th December	Dry and cold	Dry and cold
Pauṣa	15th December- 14th January	Snow and cold	Dry and cold

Source: Field data.

3.2.6 Impact of climate change on day to day lives

The daily activities were perceived to be affected differently by female and male respondents. The female respondents felt that the decline in rainfall and snowfall had led to water scarcity which was a major issue as water was needed in all the household chores, be it cooking or cleaning. Not only the changing climate have affected the respondents' day to day activities, it has also caused additional stress on them. The unseasonal and untimely rainfall was affecting their daily activities. The increasing temperature and random lashes of rain was interrupting their work in the field.

A respondent (42-year-old, female, from lower altitude village, Bhawana) stated that:

> *Pānī kī baṛī dikkata hai. Ghara ke kāma ke liye to pānī chāhiye hī, auraton ko aur bhī kāmo ke liye pānī chāhiye hotā. Jab kapaṛe āte hai to aur mushakil ho jātī hai. Pānī bahuta lagatā hai usa* time, *khuda ke liye, phira sāre kapaṛe dhonā, saphāī karanā. Ādamiyo kā kyā hai, pūrā din bāhara rahate.*

> There is water problem. Water is required for household chores, but women need water for other purpose as well. It becomes more difficult during the periods. More water is used at that time, for oneself, then all the clothes have to be washed, for cleaning (from both hygiene and cultural point of view). The men are not bothered, they stay outside the whole day.

Another respondent (in her late 80s, female, from lower middle altitude village, Kiari) complained:

> *Mausama sātha nahīn de rahā to seba utanā lagatā nahīn. Bachche bhī Shimla me hī hote hai pūrā sāla,* season time *para āte hai basa. Aba akele kitanā kāma kara lungī. Bāriśe hotī nahīn, pānī bachā nahīn bāoṛiyon me.* Lift *kā pānī 14-15 dinon bāda ātā hai. Āsa pāsa ke bachcho se mangavātī hu pānī zarūrata paṛane para.*

> As the climate is not helping, apple production has declined. Children stay in Shimla the whole year, they only come during season time (harvesting season). Now how much work will I get done alone. It does not rain, there is no water in *bāoṛiyān* (water source). Lift water is supplied after 14-15 days. I ask the children around to bring water when needed.

A respondent (79-year-old, female, from lower-middle altitude village, Kiari), was saddened that the women had to suffer more. She responded that:

> *Pānī ke binā roza ke kāma sambhālanā muśkil to hai hī, pati aur bachcho kī ḍānṭa bhī sunanī paṛatī hai. Inako bhī* tension *rahatī hai. Binā barpha aura bāriśon ke bāgīche sūkha rahe hai. Uskā gussā hama para nikālate hai.*
>
> It is difficult to handle everyday work without water, I also have to listen to scolding of husband and children. They also have tension. The orchards are drying without snowfall and rainfall. They take out that anger on us.

Women felt random lashes of rain during summer have increased their daily load of work. The respondent (48-year-old, female, from the upper-middle altitude village, Kufar Bag) told that:

> *Hamāra kāma* double *ho gayā hai bemausami bāriś se. Kapaṛe dho kara bāhara lagāye hote hai sūkhne ke liye ke achānaka bāriśa ā jātī. Saba kapaṛe phira se dhone paṛate. Pānī kī tangī vaise hī ho rahī hai. Paraso jaise baṛiyā dālī thī chhatta para sūkhne ke liye, sāpha āsamāna thā. Achānaka se bādala āye aur bāriś se sārī baṛiyā kharāba ho gaī.*
>
> Our work has doubled due to unseasonal rainfall. The clothes are hanged outside for drying and suddenly the rain comes. All clothes have to be washed again. The water is already getting scarce. Few days back I

had kept *baṛiyā* (small ball of *Vigna mungo*, the black gram) on roof for drying, the sky was clear. Suddenly, the clouds came and all the *baṛiyā* were ruined by rain.

A respondent (59-year-old, female, from the higher altitude village, Jashla) explained that:

> *Abhī itanī dikkata nahīn hui hai hamen. Pānī kā guzārā ho jātā hai. Para garmī bahuta baṛa gaī hai aba yahan. Bāhara kāma karanā muśkil ho jātā hai kabhī kabhī. Garmī hai par yahān havā ṭhandī chalatī hai. To itanā koi pharka nahīn paṛā hai roza ke kāmo para. Hān, havā nahīn chalegī to muśkil ho jāyegā.*
>
> We do not have so much problem right now. Water requirements are met. But the temperature has increased a lot here. It becomes difficult to work outside at times. it is hot but the wind that blows here is cold. So, the everyday work has not been affected much. Yes, if the wind does not blow, then it will get difficult.

The recent events of untimely rainfall during the apple harvesting season, August-September, 2021, has become a separate issue for the women of the field area. As per a respondent (54-year-old, female, from the upper-middle altitude village, Kufar Bag):

> *Isa sāla to būre hi hāl hai bāriśon ke.* Season time *mein kaba hotī itanī bāriś? Ghara me phīlen ā rahī, bāhara kīchaṛa hī kīchaṛa ho rakhā hai. Kitanī saphāī kara lungī main akele? Upara se* labour *vālo ko bhī dekhanā paṛatā hai. Unake liye chāya bānaāye ke saphaī karen? Ye pūrā dina bāhara rehate. Gorakhen aur bharānī bhī sāra kīchaṛa se gandā kara rahe. Inako bhī kyā bole, apnā kāma kara rahe. Kuchha bolenge to kāma nahīn karenge ḍhanga se.*
>
> The condition of rainfall is bad this year. When does it rain so much during season time? *Phīlen* (*Gastopoda*, slugs) are coming in the house, it is muddy outside.

> How much will I clean alone? Above that, I have to look after the labour as well. One will make tea for them or clean? He (husband) stays outside the whole day. *Gorakhen* (Nepali labour) and *bharānī* (worker that sort and fills apple in the boxes) also make mess with mud. What to say to them, they are doing their work. If we'll say anything, they will not work properly.

We can deduce from the above given narratives that it was woman who faced the brunt of climate change on daily basis. The male respondents did feel that climate change was affecting the daily activities like their work in the fields, for instance, repetition of spray in case of untimely rainfall, or increase in efforts required on daily level to keep up with the diseases and pests arising in the orchards so as to sustain the output of their livelihood. However, as they spent most of the time in orchards, they did not feel the equal amount of burden that their female counterparts felt. Majority of the male respondents, being the household heads in the patriarchal society, were more concerned about the long-term impact of climate change on various natural resources and the mode of subsistence, horticulture, that was dependent on the climate and natural resources. The study by Oduniyi (2014) in Mpumalanga province, South Africa, found that climate change affected individuals and groups differently, for instance, the elders, children, tribal group, and low-income population were more sensitive to climate change impact than others. This study is consistent with the finding of the present study that the day-to-day activities of women were affected more than that of men in the study area.

3.3 Comparison of people's perception and climate data

It was found during the analysis of the data that both respondents' perception and the climate data points toward an increasing trend in temperature. The perception of people about rainfall pattern being erratic has been supported by the climate data for rainfall for past 50 years. The respondents' observation about overall decline in rainfall is similar to the trend shown by rainfall for past five decades. A decline in summer rain or pre-monsoon rainfall was observed by the respondents of the field area, which confers with the data provided by the Meteorological Centre (IMD), Shimla.

The linear regression trendline presents an increasing trend for the post-monsoon rainfall and a decreasing trend for monsoon rainfall. These results support the perception of people regarding shift in the months of monsoon rainfall from July-August towards August-September. However, the trend in snowfall based on climate data provided by IMD does not match the perception of respondents. According to the respondents, snowfall has observed a drastic change, both in amount and in seasonality (timings).

The increasing trend presented by the statistical analysis is based on the data of eleven years alone and it would not be out of place to mention that the amount of snowfall in the year 2020 was maximum in the decade which may have affected the trend. Also, the statistical data is cumulative in nature and gives a general trend based on the recorded data. However, one can attribute the decline in snowfall to the warmer winter, as perceived by the respondents and inferred by the recorded mean minimum temperature that has shown a sharp increase in the past decade. The extent of climatic parameters' effect has shown variability according to the altitude, based on the data collected by the researcher from the field area.

4

Impact of Climate Change on Crop Varieties, Calendar and Crop Production

This chapter deals with documentation of crop varieties, crop calendar and crop production in the areas under study. The third objective, i.e., to document crop varieties, crop production, their cropping time and their shifting calendar due to climate change has been covered under this chapter by taking into consideration both, people's perception and the data provided by the Department of Horticulture, Shimla, Himachal Pradesh. The livelihood of the people of study area is mainly based on horticulture. According to *The Oxford English Dictionary* (1989), "Horticulture is the art or science of cultivating or managing gardens, including the growing of flowers, fruits, and vegetables." As per *The New Encyclopaedia Britannica* (2002), "Horticulture is a branch of agriculture concerned with the cultivation of garden crops, generally fruits, vegetables, flowers, and ornamentals such as plants used for landscaping".

Horticulture requires some particular conditions, special techniques and equipments. The knowledge of the horticulturist and the success of the crop are influenced by many factors, namely, climate, terrain, and other regional variations. The fruits grown in the study area needs proper chilling hours during winter dormant phase, pruning and spray at regular intervals. The temperature, snowfall and rainfall are crucial for the success of any fruit production. The factors like altitude and whether the area is sun facing or shady also impact the horticulture practice.

4.1 Crop documentation

Kotkhai tehsil have different varieties of crops being grown at different altitudes. While some vegetables are also grown, they are not for sale purpose and are grown for gardening purposes, they have not been documented under crop, as crop is something cultivated for sale purpose or subsistence. Kotkhai tehsil is famous for its production of apple which is grown all over the Kotkhai region. Pome fruits like apple (*Malus pumila*) and pear (*Pyrus communis*), and stone fruits like plum (*Prunus domestica*), peach (*Prunus persica*), apricot (*Prunus armeniaca*) and cherry (*Prunus avium*) have been documented in the field area during the present research (Table 4.1). Apple comprises the bulk of the horticultural produce in the field area. In the villages Bhawana, Kiari and Jashla, people cultivate both pome and stone fruits, that are, apple, pears, cherry, plum, peach and apricot. While in Kufar Bag people only cultivate pome fruit, apple.

Table 4.1: List of crops grown for commercial purpose

Name of the villages	Name of the crops
Bhawana	Apple, pears, plum, peach, apricot and cherry
Kiari	Apple, pears, plum, peach, apricot and cherry
Kufar Bag	Apple
Jashla	Apple, pears, plum, peach, apricot and cherry

Source: Field data.

4.1.1 Evolution of the present horticulture practices

As the human civilization evolved during the course of time, people around the world developed means of livelihood best suited to the climate of the place they inhabited (Kulkarni and Leary, 2007). In convergence to this view, earlier the people of Kotkhai tehsil used to grow potato (*Solanum tuberosum*), *bajra*/ pearl millet (*Pennisetum glaucum*) and had a non-vegetarian diet. This was mainly due to the fact that there were no transport facilities and the climate was harsh, and it was reasonable to grow these crops in such conditions and create a self-sustaining mode of subsistence, and also, the meat could be stored easily during long winter spells.

Over the time, people of Kotkhai tehsil shifted the means of livelihood to apple production, a cash crop, which was sustainable in the type of environment Kotkhai had. With time it has emerged as a major apple producing belt of Himachal Pradesh. The cash crop means of livelihood has been assisted by the development of roads, which was primarily responsible for deforestation and consequently climate change. Also, with development of roads, it became easier to access markets and procure material for day-to-day activities and requirements, and thus a need to carry on with self-sustaining mode of subsistence waned.

But as the climate is changeable, non-conformities due to change can be unsettling (Toulmin *et al.*, 2000). The disruptions due to the anthropogenic climate change has become a challenge today (Weart, 2004). The people of Kotkhai residing in villages at different altitudes have suffered the brunt of changing climate at a different level. During the research it has been observed that due to changing pattern of precipitation, and decreasing rainfall and increasing temperature, the villagers have been dealing with a major decline in apple economy at lower and lower middle altitude villages, viz., Bhawana (1825 m) and Kiari (2224 m). However, people in villages like Kufar Bag (2362 m) and Jashla (2438 m), the upper middle altitude and high-altitude villages respectively, have not faced dire consequences as Bhawana and Kiari, but are facing troubles due to erratic rainfall and increasing hail precipitation during recent times, which is impacting the apple crop. New varieties of apple along with other stone fruits have started being grown in the lower and lower-middle altitude villages of Bhawana and Kiari over the time. Currently, pears, plum, peaches, apricot, pear and cherry are also being grown along with apple in the field area.

4.1.2 Changes in crop varieties with changing climate

Climate change have affected the type of fruit crops being

cultivated the field area. New varieties of crop are being produced along with the traditional varieties (Table 4.2). According to the respondents there are multiple reasons for advent of new varieties in the area, climate change being one of the major factors. A respondent (51-year-old, male, from lower-middle altitude village, Kiari) stated regarding apple varieties that,

> *Pahale purāne wāle paudhe 60-70 sāla chala paṛate the ārāma se. Aba 30-40 sāla bāda paudhā sūkha rahā hai,* fruit size *bhī kama ho rahā hai. Vajaha ye hai kī pahale mausama achchha thā, bāriśe* time *para hotī thī, barpha bhī khūba paṛatī thī. Aba namī kī kamī ke kāraṇa sūkhā paṛa rahā hai. Hame aise paudhe chāhiye jo jaladī* production start *kare. Puranī varieties 10 sāla taka* production *detī hai achchhī, para ye jo* new varieties *hai jaise Gala aura Spur, ye* dwarf variety *hai, 2-3 sāla taka* mature *ho jātī hai aura* production start *kara detī hai. Abhī* Gala *kī* market *bhī achchhi hai aura* instant return *kā zamānā hai to phāyadā bhī zyādā hai ina varieties kā.*
>
> Older/traditional variety trees used to bear fruits for about 60-70 years before. Now after 30-40 years the plant is drying out, fruit size is also decreasing. The reason being that earlier the climate was nice, it used to rain on time, snow a lot. Now due to lack of moisture they are drying out. We need plants that start production early. Older varieties took 10 years to start production, but the new varieties like Gala and Spur, the dwarf varieties, mature within 2-3 years and start production. Right now, the market is good for Gala and as the time is about instant returns, the profit is more in new varieties.

Above given narrative points at the need to use new varieties as they are better adapted for the changing climate. The findings of the study suggested that climate change have affected the type of fruit crops that are being cultivated at the field area. New varieties of crop are being produced along with the traditional varieties.

However, all new varieties of crop are not being favoured over the traditional varieties. For instance, according to the respondents, the new smooth skin variety of peach is not being favoured as much as the traditional brushed variety and majority of the people in Kufar Bag cultivate the traditional variety of apple.

Table 4.2: Traditional and new varieties of crop documented in the field area

Crop	Traditional variety	New variety
Apple	Royal Delicious, Red Delicious, Rich-a-red, Red Gold, Golden, King Pipin, Granny Smith	Spur varieties – Oregon I and II, Scarlet I and II, Super Chief Gala varieties – Gala Gala, Redlum Gala, Dark Baron Gala, Fenplus Gala
Pear	*Moti Dandi* (Red Bartlett), Bargument	Pekham, Karmin
Plum	Santa Rosa, Mariposa	Red Beaut, Black Amber, Prune
Peach	Brushed	Smooth skin
Apricot	*Śakarapārā*	-
Cherry	Black cherry	-

Source: Field data.

4.2 Crop calendar

The crop calendar is dependent on the altitude of the area, amount and time of snowfall as well as rainfall, and whether the temperature requirements are being met. The respondents found that the flowering time may be early or same, but the harvesting time was getting delayed by almost a week or so due to factors like delay in snowfall, monsoons and unfavourable temperature conditions which led to increase in maturation period. People's perception about the chilling period is portrayed in Figure 4.1. Out of the total respondents from the selected villages, 49 per cent noted a decrease in chilling period due to declining snowfall, 37 per cent observed the chilling period to be non-uniform due to variation in timing and amount of snowfall, and 14 per cent noticed the chilling period to be the same over the years.

People's perception regarding flowering time is depicted in Figure 4.2. It is evident from the figure that about 48 per cent of the total respondents responded that the flowering started earlier while 52 per cent observed it to be the same.

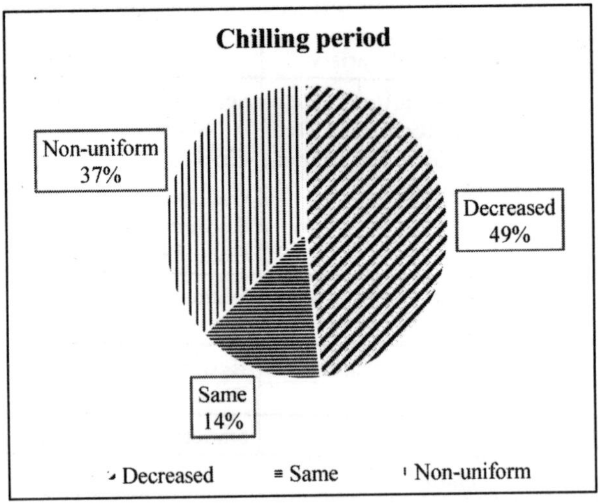

Figure 4.1: People's perception about the chilling period in the study area.

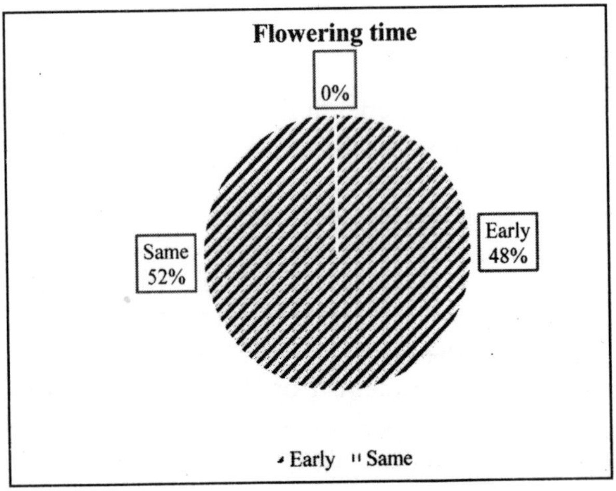

Figure 4.2: People's Perception regarding flowering time in the study area.

People's perception about the harvesting time has been portrayed in Figure 4.3. All the respondents found a delay in harvesting time regardless of the flowering time.

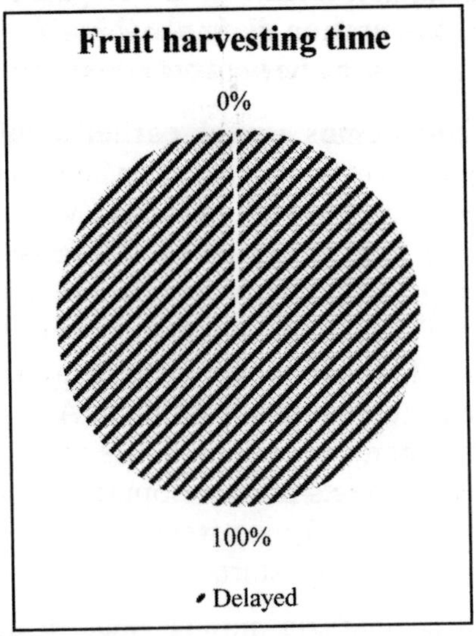

Figure 4.3: People's perception regarding harvesting time in the study area.

The respondents at lower altitude and lower-middle altitude villages of Bhawana and Kiari observed a trend of early flowering in plants and a delay in harvesting time due to negligible amount of snowfall in recent years. A respondent (62-year-old, male, from the lower altitude village, Bhawana) observed that,

> Flowering *eka* week early *ho gaī hai aura* development *eka* week late.
> (Flowering occurs one week earlier and development one week late.)

Another respondent (84-year-old, male, from the lower-middle altitude village, Kiari) perceived that,

> Flowering *aba eka haphtā pahale hī ho rahī. Bākī* fruit development temperature *kaisā hai, kītanī barpha paṛī, kaba paṛī, bāriś kitanī huī, ina saba para nirbhara karatī. Agara barpha paṛī* timely *to* timely flowering *hotī. Agara* late *ho, jaise Jan-Feb kī jagaha Mārcha*

> *me to* flowering *to ho jātī apne* time *pe para bākī* development late *ho jātī kyonki* heat requirement *pūrī nahīn hotī. Aura agara barpha nā paṛe to* chilling hours *pūre nahīn hote to usse bhī* development *para pharka paṛatā.* Maturation *ek haphatā late ho gayī hai jisa vajaha se* season *aba haphatā* late start *hotā.*
>
> Flowering begins a week earlier now. Rest of the development of fruit depends on how the temperature is, how much snowfall occurred, when it occurred, how much rainfall happened. If it snows timely, then flowering will occur timely. If late, like instead of January-February in March, then flowering is on time but rest of the development is delayed because the heat requirement is not fulfilled. And if there is no snowfall then the chilling hours are not completed which also affects the development. The maturation has been delayed by a week because of which the season (harvesting) starts late.

While in upper-middle altitude village, Kufar Bag, flowering remained unaffected and in higher altitude village, Jashla, the flowering and setting depended on the timing of the snowfall and temperature conditions. Late snowfall resulted in late flowering and delay in overall process of development of the fruit which delayed the harvesting time. Also, the development process was slower in higher regions due to lower temperature which affected the growth/size of the fruit. A respondent (59-year-old, male, from the upper-middle altitude village, Kufar Bag) believed that,

> Setting *pahale* height *me ho jātī hai para dānā pahale* lower height *me baṛā hotā hai kyonki jaba garmī me bāriś paṛatī hai to* sap lower height *me jaladī caṛatā hai jisa se* size *jaldī baṛhatā. Zyādā* height *para ṭhanda zyādā paṛatī to* size *bhī* late *bānaatā hai.*

The fruit setting finishes earlier in higher regions but the fruit size increases first in lower height area because during summer when it rains, the sap rises faster because of which the size increases earlier. It is colder at higher altitude so the increase in size is late.

Table 4.3 Changes in crop calendar (apple and pear) from green tip stage to harvesting time observed in the selected villages

Phase	Villages							
	Bhawana		Kiari		Kufar Bag		Jashla	
	Before	Now	Before	Now	Before	Now	Before	Now
Green tip	After 15th March	Before 15th March	After 15th March	Before 15th March	Till last week of March	Till last week of March	Till last week of March	Till last week of March
Bud	5-6 days later	5-6 days later	5-6 days later	5-6 days later	5-6 days later	5-6 days later	5-6 days later	5-6 days later
Flowering	First week of April	End of March	First week of April	End of march	1st -14th April	1st -14th April	1st -14th April	15th -20th April
Setting	A week after flowering	A week after flowering	A week after flowering	A week after flowering	A week after flowering	A week after flowering	A week after flowering	A week after flowering
Petal fall	April end	April end	April end	April end	May	May	May	May
Fruit Development	May	May-June	May	May-June	June	June-July	June	June-July
Fruit Ripening	June	June-July	June	June-July	July-Mid August	July-August	July-August	July-1st week of September
Harvesting	Mid July	July end	Mid July	1st August	Mid-August onwards	1st week of September	1st week of September	10th-15th September

Source: Field data.

Table 4.4: Changes in crop calendar (plum and cherry) from green tip and bud stage to harvesting time observed in the selected villages

Phase	Villages							
	Bhawana		Kiari		Kufar Bag*		Jashla	
	Before	Now	Before	Now	Before	Now	Before	Now
Green tip and bud	Mid-February to February end	Mid-February to February end	Mid-February to February end	Mid-February to February end	-	-	March beginning	March beginning
Flowering	1st week of March	1st week of March	1st week of March	1st week of March	-	-	Mid-March	Mid-March
Setting	March	March	March	March	-	-	March	March
Petal fall	March end	March end	March end	March end	-	-	April	April
Fruit Development	April	April	April	April	-	-	May	May
Fruit Ripening	Mid-May	May	Mid-May	May	-	-	June	June
Harvesting	May end	June	May end	June	-	-	July end - August	July end-August

Source: Field data.

**plum and cherry are not cultivated in Kufar Bag.*

Table 4.5: Changes in crop calendar (peach and apricot) from green tip and bud stage to harvesting time observed in the selected villages

Phase	Villages							
	Bhawana		Kiari		Kufar Bag*		Jashla	
	Before	Now	Before	Now	Before	Now	Before	Now
Green tip and bud	Last week of February	Last 14 days of February	Last week of February	Last 14 days of February	-	-	March	March
Flowering	March	March	March	March	-	-	April	April
Setting	March-April	March-April	March-April	March-April	-	-	April end	April end
Petal fall	April	April	April	April	-	-	May	May
Fruit Development	May	May	May	May	-	-	June	June
Fruit Ripening	June	June end	June	June end	-	-	June-July	June-July
Harvesting	June end	July	June end	July	-	-	July end-August	July end-August

Source: Field data.

**peach and apricot are not cultivated in Kufar Bag.*

A comparative account of the variations observed in crop calendar from green tip and bud stage to harvesting time by the respondents in all the four villages is presented from Table 4.3 to Table 4.5. Table 4.3 presents the changes in different stages of growth from green tip till harvesting for apple and pear in the selected villages as observed by the respondents. The stages from green tip to setting, especially flowering has depicted a trend of early onset, by almost a week. While fruit development, ripening and harvesting have been delayed by a fortnight in all the selected villages.

Table 4.4 demonstrates the changes in crop calendar for plum and cherry in the villages of Bhawana, Kiari and Jashla (plum and cherry are not cultivated in Kufar Bag). The respondents did not notice any change in the stages from green tip to fruit development. A delay has been documented in case of fruit ripening and harvesting of the crop by approximately two weeks in the lower altitude village of Bhawana and lower-middle altitude village of Kiari. The high-altitude village of Jashla did not present any changes in crop calendar for plum and cherry.

As per the Table 4.5, the green tip and budding has been early by a week, and fruit ripening and harvesting have been delayed by almost two weeks for lower altitude village of Bhawana and lower-middle altitude village of Kiari. The stages from flowering to fruit development has not depicted any changes for the villages of Bhawana and Kiari. Also, the crop of peach and apricot did not display any changes in crop calendar for the high-altitude village of Jashla.

Changes in pre-flowering stages, flowering, fruit set and fruit development stage, fruit growth and development stage, and harvesting time have been observed in the field area mainly due to change in climatic parameters over the time and also due to inconsistent pattern occurring due to climate change. For instance, the recent case of snowfall in the month of April, 2021 has resulted in further delay in the harvesting of apple. Already a delay in harvesting was observed by the respondents from different altitude villages. The untimely snowfall in April not only damaged the crop, but also extended the already delayed harvesting phase.

According to a respondent (51-year-old, male, from the lower altitude village, Bhawana),

> *Pahale* season 15th July *taka* start *ho jātā thā.* 2010 *se lagabhaga* July end *me shurū ho rahā hai. Isa sāla* (2021), August *taka* shift *ho jāyegā.*
>
> Earlier season used to start around 15th July. Since 2010, it starts around July end. This year (2021), it will shift to August.

Another respondent (37-year-old, male, from the higher altitude village, Jashla) have noticed a shift in the harvesting time,

> *Pahale tuṛāna* September *ke pahale* week *me* start *hotā thā. Ab* 10th September *taka hotā hai.* But *isa sāla* (2021), 15th September *taka shurū karenge kyonki* size *hī nahīn bānaā, nā ranga ā rahā hai.*
>
> Earlier apple picking began in first week of September. Now it starts by 10th September. But this year (2021), will start by 15th September because the size has not increased and there is no colour.

Similarly, there has been a shift in harvesting of apple from 20th August to 10th September over the time in Kufar Bag. The changes in various growth phases and harvesting time is in convergence to the study by Ladaniya (2011), according to which, increasing temperature is affecting flowering and fruit development in Himachal Pradesh, and changed heat summation would affect fruit development and duration of maturity.

4.3 Crop production

Both people's perception and data provided by the Department of Horticulture (Shimla) were taken into account while analysing the crop production.

4.3.1 People's perception

Respondents from different altitude villages had different perception regarding the overall production and productivity o the crop grown. While the productivity was perceived to decli over the time by all the respondents, the reason and net amount of production varied from village to village. As per the respondents, apple, the main crop being produced in the area, is shifting from lower altitude to upper altitude areas due to unfavourable climate conditions in lower regions. The stone fruits are however, doing fine at lower altitude areas. As per a respondent (56-year-old, male, from the lower altitude village, Bhawana),

> *Pichhale* 10 *kuchha sālo se barpha nahīn paṛa rahī ḍhanga se. Bagīche sūkha rahe hai. Seba ke liye barpha paṛanā baṛā jarūrī hai. Seba kī* chilling requirement *pūrī nahīn ho rahī kyonki aba sardiyon me bhī tāpamāna baṛha gayā hai. Pahale jaisī ṭhanda nahīn. Pichhale eka do sāla se to barpha paṛī hī nahīn hai samajho,* 1-2 inch *hī paṛī hogī. Aise me seba ka* size *nahīn bāna rahā. Bāriśe bhī kama ho gaī hai.* Monsoon time *para nahīn āta. Garmī me bhī jo bāriś hotī thī, jise* local rains *kahate hai, vo bhī nahīn ho rahī. Sūkha paṛa rahā hai. Achchhi phasala ke liye jo* balance *chāhiye vo kharāba ho gayā hai. Aba* Italy *se naye* saplings *lāye hai* high density apple plant *ke. Inakī* chilling requirements *kama hai. Dekhate hai* 2-3 *sāla me kaisā* response *dete hai. Bākī aba* stone fruits *thoṛe zyādā lagā diye hai. Garama mausama kā una para koi zyādā* effect *nahīn hotā.*
>
> From past 10 years the snowfall is not proper. Orchards are drying, snowfall is crucial for apple. The chilling requirement of apple is not be fulfilled because now the temperature is rising in winter. It is not as cold as before. It has not snowed for past one two years, only 1-2 inches. In such situation, the size of apple is not developing. Rainfall have declined. Monsoon does not arrive on time. Rainfall in summer, the local rains, is not happening. There is drought like conditions. Balance needed for a good crop has been impaired. Now, high density apple plants' sapling has been brought from Italy. These have a lower chilling requirement. Let us see what response they give in 2-3 years. Rest, more stone fruits are being grown. Warmer climate does not affect them more.

Another respondent (52-year-old male, from lower-middle altitude village, Kiari), observed that,

> Plum, cherry, *śakarapāre to hama bachapana se hī dekha rahe hai bagīcho me. Una para zyādā asara nahīn huā hai mausama kā. Para purāne seba ke peṛa aba sūkha rahe aura naye lagāne ke liye namī nahīn rahī hai. Kāraṇa ye kī aba barpha nahīn pāṛatī pahale jaisī. Barpha se sabase zyādā namī milatī zamīna ko kyonki vo achhe se zamīna me rachatī hai aura* groundwater *ko* recharge *karatī hai. Bāriśe jo ho rahī hai vo to basa baha hī jātī hai.*
>
> We have been seeing plum, cherry and apricot in orchards since childhood. There is not much effect of climate on them. But the old apple trees are drying out and there is no moisture left for new trees. Reason being that now it does not snow as before. Snow provides maximum moisture to the land as it percolates nicely and recharges the groundwater. Rainfall that is occurring just drains away.

A respondent (73-year-old, male, from the upper-middle altitude village, Kufar Bag), felt that there was no drastic difference in production in the village. According to him,

> *Seba to aba pahale se zyādā hote hai. Aba naye peṛa bhī baṛe ho gaye hai to unakī* production *bhī shurū ho gaī hai. Hāla ye hai ke eka eka peṛa kī* production *kama ho gaī hai agara* individually *dekhe. Pahale peṛa para* uniform *dānā lagatā thā. Aba kahīn ṭahaniyon para zyādā dāne hai kahīn kama. Usa se* overall growth *para pharka paṛa jātā hai. Hama shīle me hai to hamen zyādā nukasāna nahīn hai* temperature *baṛhane ka kyonki barpha dera se hī pighalatī hai.*
>
> Apple production is more than before. Now new trees have matured so their production has also started. Yes, each trees' individual productivity has declined. Earlier apples were uniformly distributed on tree. Now some branches have more apples and some have less. This

> affects the overall growth. We are in the shade, north facing slope, so we are not at the disadvantage due to increasing temperature as the snow melts late.

It can be concluded from the above-mentioned narratives that an increase in the production of apple may be attributed to increase in the area, as new varieties of crop were being grown which gave good results within small period of time, and the more tress have matured over the period of time. The result is in convergence to the study of Ahmed *et al.* (2011), who stated that in spite of increase in temperature, the hill states have shown an increase in the area and production of apple. Also at higher altitude, as the weather was getting warmer, the areas that used to stay under snow cover and were not suitable for any crop production are now turning into suitable options for production of apple. The respondent (61-year-old, male, from the high-altitude village, Jashla) stated,

> *Aba to ūpara* Deorighat *me bhī logon ne seba lagāne shurū kara diye hai. Garmī baṛha gayī hai to aba barpha zyādā dera taka nahīn ṭikatī. Pahale yahan jaba* flowering *hotī thī taba bhī ūpara barpha rehatī thī. Aba to itanī barpha hī nī paṛatī. Jo paṛatī hai vo bhī jaladī pighala jātī.*

> Now even above in Deorighat (a place at a higher altitude than Jashla), people have started apple cultivation. As the temperature has increased so the snow does not stay for long. Earlier when the flowering started here, the snow was still there above. Now it does not even snow so much. Whatever snow melts away quickly.

From the above narratives it can also be inferred that the snowfall plays the most important role in apple production. Pear crop require almost similar climatic conditions as apple, therefore, the productivity shows similar pattern as apple according to the respondents. The warmer temperature favours stone fruit development as they require fewer chilling hours, therefore people of lower and lower-middle altitude have shifted to stone fruit production. The shift in the varieties have been supported by the study of Ahmed *et al.* (2011), in which they proposed a shift from high chill varieties to varieties that required low chilling hours, i.e., new types of apples (High Density Plant), peaches, etc. The production has been different for different altitude villages, but the productivity has been declining at lower altitude areas and is non-uniform at higher regions.

4.3.2 Crop production based on secondary data

According to the data provided by the Department of Horticulture (Shimla) for Kotkhai tehsil (2007-2017), apple is the most widely grown fruit crop in the area and comprises the largest portion of the crop being produced in the Kotkhai belt, followed by pear, cherry, plum, apricot and peach.

Table 4.6 gives the record of production (MT) and area of production (ha) and productivity (MT/ha) of apple in Kotkhai tehsil. As per this data, an increase in area has been observed, while the production and productivity has shown an irregular pattern throughout the years, i.e., 2007-2017. The scatter chart (Figure 4.4) showing the trend of productivity over the years (2007-2017) indicate towards maximum productivity in year 2011 and minimum in year 2012.

Table 4.6: Apple production (MT), production area (ha) and productivity (MT/ha) in Kotkhai tehsil

Year	Apple production (MT)	Area under apple cultivation (ha)	Productivity (MT/ha)
2007	65320	6143.59	10.63
2008	127150	6143.59	20.69
2009	131940	6300.77	20.94
2010	70536	6506.4	10.84
2011	**198800**	**6718.32**	**29.59**
2012	**55616.16**	**7027.86**	**7.91**
2013	85719.79	7408.57	11.57
2014	143334.11	7440.85	19.26

(Contd.)

2015	114904	7642.23	15.04
2016	143400	7867	18.23
2017	90564	7867	11.51

Source: Department of Horticulture (Shimla).

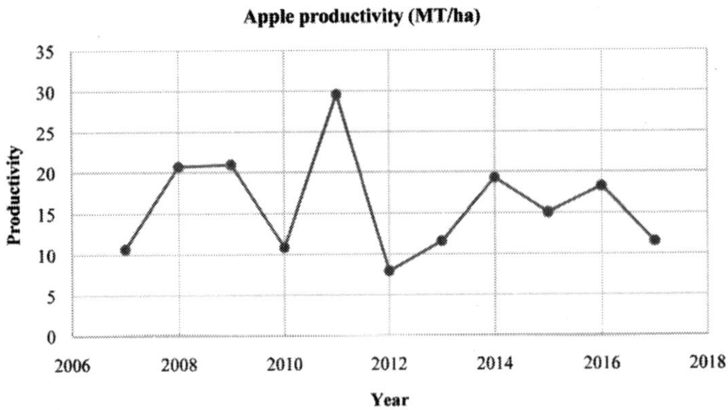

Source: Based on data from Department of Horticulture (Shimla).

Figure 4.4: Scatter chart showing apple productivity (MT/ha).

The productivity of apple is dependent on climatic conditions like temperature, rainfall and snowfall. Table 4.7 shows the climatic conditions during different growth phase of the fruit for the years 2007-2017 for temperature and rainfall, and years 2010-2017 for snowfall. During the year of maximum productivity (2011), the recorded mean of snowfall and summer rainfall is highest while the recorded mean for same is lowest for the year of minimum productivity (2012). This result synchronizes with the perception of the people that the snowfall and summer rains are important for apple productivity.

Pear is the second most grown fruit crop in the Tehsil Kotkhai. Table 4.8 gives the record of production (MT), area of production (ha), and productivity (MT/ha) of pear in Kotkhai tehsil. As per this data, slight increase in area has been observed, while the production has been constant for the years 2007-2009 followed by a drastic increase in production in year 2010. After 2010 there has been a random decline in production. The productivity has

Table 4.7: Productivity and climate conditions during different growth phases of apple from year 2007 to year 2017

Year	Prod-uctivity (MT/ha)	Dormancy/Pre-flowering stage (January-March)				Flowering, fruit-set and fruit development stage (April-June)				Fruit growth and development stage (July-September)			
		Max temp (°C)	Min Temp (°C)	Rain-fall (mm)	Snow-fall (cm)	Max Temp (°C)	Min Temp (°C)	Rain-fall (mm)	Snow-fall (cm)	Max Temp (°C)	Min Temp (°C)	Rain-fall (mm)	Snow-fall (cm)
2007	10.63	13.6	4.7	92.6	NA	24.4	15.4	20.1	0	22.6	15.5	131.0	0
2008	20.69	13.9	4.9	0.0	NA	22.6	13.9	0	0	22.2	15.5	211.7	0
2009	20.94	16.2	6.9	24.0	NA	24.8	14.7	31.0	0	23.0	16.2	189.1	0
2010	10.84	17.1	7.2	46.7	10.8	26.3	16.2	72.9	0	22.6	16.0	218.3	0
2011	**29.59**	**14.9**	**5.3**	**NA**	**29.6**	**23.8**	**13.8**	**105.4**	**0**	**22.6**	**16.0**	**150.8**	**0**
2012	**7.91**	**13.7**	**4.2**	**63.3**	**7.9**	**25.1**	**14.8**	**4.7**	**0**	**22.7**	**16.3**	**110.6**	**0**
2013	11.57	14.6	5.2	120.2	11.6	24.1	15.1	75.0	0	23.1	16.2	74.8	0
2014	19.26	14.0	4.1	146.3	19.3	24.6	13.9	87.8	0	24.0	15.7	132.9	0
2015	15.04	14.3	5.3	122.4	15.0	23.9	14.4	29.4	0	23.2	15.8	191.8	0
2016	18.23	16.1	6.1	92.1	18.2	25.1	15.2	61.2	0	23.2	15.9	96.0	0
2017	11.51	15.2	5.7	79.4	11.5	24.4	14.8	128.8	0	23.2	16.1	141.0	0

Source: Based on data provided by IMD and Department of Horticulture (Shimla).

shown an erratic pattern throughout the years, 2007-2017. The scatter chart (Figure 4.5) showing the trend of productivity over the years (2007-2017) indicate towards maximum productivity in the year 2010 and minimum in the year 2009.

Table 4.8: Pear production (MT), production area (ha) and productivity (MT/ha) in Kotkhai tehsil

Year	Pear production (MT)	Area under pear production (ha)	Productivity (MT/ha)
2007	675	456.4	1.48
2008	675	456.4	1.48
2009	**675**	**468.78**	**1.44**
2010	**6633**	**482.96**	**13.73**
2011	5870	489.81	11.98
2012	1611.1	493.24	3.27
2013	4750	492.24	9.65
2014	4696	491.24	9.56
2015	3000	490.01	6.12
2016	5700	493	11.56
2017	3900	493	7.91

Source: Department of Horticulture (Shimla).

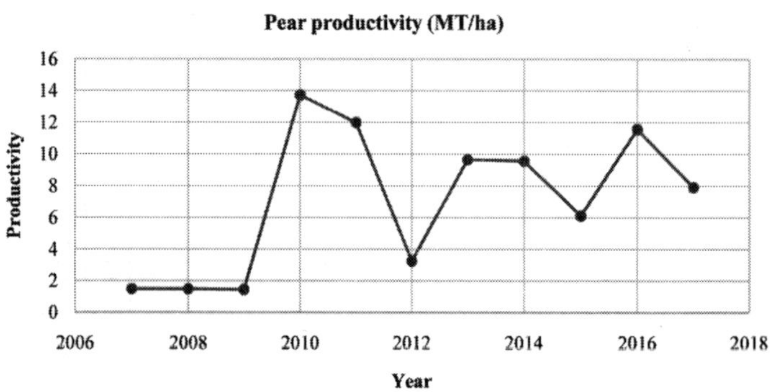

Source: Based on data from Department of Horticulture (Shimla).

Figure 4.5: Scatter chart showing pear productivity (MT/ha).

Table 4.9: Productivity and climate conditions during different growth phases of pear from year 2007 to year 2017

Year	Prod-uctivity (MT/ha)	Dormancy/Pre-flowering stage (January-March)				Flowering, fruit-set and fruit development stage (April-June)				Fruit growth and development stage (July-September)			
		Max temp (°C)	Min Temp (°C)	Rain-fall (mm)	Snow-fall (cm)	Max Temp (°C)	Min Temp (°C)	Rain-fall (mm)	Snow-fall (cm)	Max Temp (°C)	Min Temp (°C)	Rain-fall (mm)	Snow-fall (cm)
2007	1.48	13.6	4.7	92.6	NA	24.4	15.4	20.1	0	22.6	15.5	131.0	0
2008	1.48	13.9	4.9	0.0	NA	22.6	13.9	0	0	22.2	15.5	211.7	0
2009	**1.44**	**16.2**	**6.9**	**24.0**	**NA**	**24.8**	**14.7**	**31.0**	**0**	**23.0**	**16.2**	**189.1**	**0**
2010	**13.73**	**17.1**	**7.2**	**46.7**	**10.8**	**26.3**	**16.2**	**72.9**	**0**	**22.6**	**16.0**	**218.3**	**0**
2011	11.98	14.9	5.3	NA	29.6	23.8	13.8	105.4	0	22.6	16.0	150.8	0
2012	3.27	13.7	4.2	63.3	7.9	25.1	14.8	4.7	0	22.7	16.3	110.6	0
2013	9.65	14.6	5.2	120.2	11.6	24.1	15.1	75.0	0	23.1	16.2	74.8	0
2014	9.56	14.0	4.1	146.3	19.3	24.6	13.9	87.8	0	24.0	15.7	132.9	0
2015	6.12	14.3	5.3	122.4	15.0	23.9	14.4	29.4	0	23.2	15.8	191.8	0
2016	11.56	16.1	6.1	92.1	18.2	25.1	15.2	61.2	0	23.2	15.9	96.0	0
2017	7.91	15.2	5.7	79.4	11.5	24.4	14.8	128.8	0	23.2	16.1	141.0	0

Source: Based on data provided by IMD and Department of Horticulture (Shimla).

Table 4.9 shows the climatic conditions during different growth phase of pear for the years 2007-2017 for temperature and rainfall, and years 2010-2017 for snowfall. The data regarding the climatic conditions during the years of minimum and maximum productivity, i.e., 2009 and 2010 respectively, does not explain the difference in the rate of production when compared with the overall data of climatic conditions.

Table 4.10 gives the record of production (MT), area of production (ha) and productivity (MT/ha) of plum in Kotkhai tehsil. As per this data, the area and production has shown a non-uniform increase and decrease throughout the years 2007-2017. The productivity has shown an irregular pattern throughout the years. The scatter chart (Figure 4.6) exhibiting the trend of productivity over the years (2007-2017) indicate towards maximum productivity in the year 2011 and minimum in the year 2017.

Table 4.10: Plum production (MT), production area (ha) and productivity (MT/ha) in Kotkhai tehsil

Year	Plum production (MT)	Area under plum production (ha)	Productivity (MT/ha)
2007	232	11.84	19.60
2008	232	11.35	20.44
2009	231	11.25	20.53
2010	261	14.81	17.62
2011	396.55	18.82	21.07
2012	273.5	19.91	13.74
2013	245	19.41	12.62
2014	317.24	18.98	16.71
2015	250	19.85	12.59
2016	294	20	14.70
2017	225	20	11.25

Source: Department of Horticulture (Shimla).

Table 4.11: Productivity and climate conditions during different growth phases of plum from the year 2007 to year 2017

Year	**Productivity (MT/ha)**	**Pre-flowering, Flowering, fruit-set stage (January-March)**				**Fruit growth and development stage (April-May)**			
		Max temp (°C)	**Min Temp (°C)**	**Rainfall (mm)**	**Snowfall (cm)**	**Max temp (°C)**	**Min Temp (°C)**	**Rainfall (mm)**	**Snowfall (cm)**
2007	19.60	13.6	4.7	92.6	NA	24.0	14.6	4.5	0
2008	20.44	13.9	4.9	0.0	NA	22.6	12.9	0	0
2009	20.53	16.2	6.9	24.0	NA	23.8	13.3	35.55	0
2010	17.62	17.1	7.2	46.7	10.8	26.6	16.0	57	0
2011	**21.07**	**14.9**	**5.3**	**NA**	**29.6**	**23.5**	**12.9**	**79.5**	**0**
2012	13.74	13.7	4.2	63.3	7.9	23.5	13.3	1.5	0
2013	12.62	14.6	5.2	120.2	11.6	23.9	14.4	41.5	0
2014	16.71	14.0	4.1	146.3	19.3	23.1	12.2	121.1	0
2015	12.59	14.3	5.3	122.4	15.0	23.2	13.7	19.7	0
2016	14.70	16.1	6.1	92.1	18.2	24.8	14.4	53.8	0
2017	**11.25**	**15.2**	**5.7**	**79.4**	**11.5**	**24.6**	**14.1**	**120.6**	**0**

Source: Based on data provided by IMD and Department of Horticulture (Shimla).

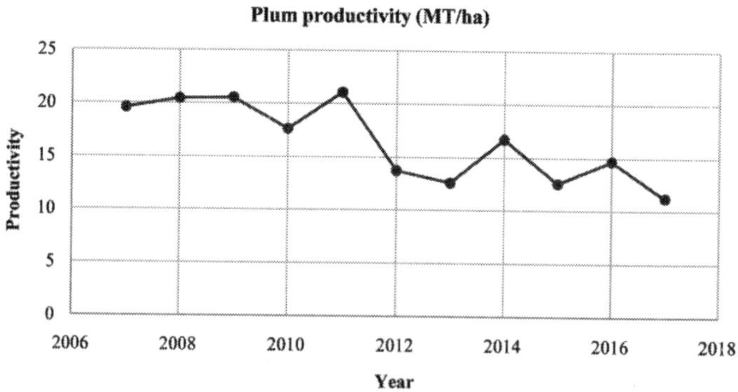

Source: Based on data from Department of Horticulture (Shimla).

Figure 4.6: Scatter chart showing plum productivity (MT/ha).

Table 4.11 depicts the climatic conditions during different growth phase of the fruit for the years 2007-2017 for temperature and rainfall, and years 2010-2017 for snowfall. The maximum snowfall for the year of maximum productivity is in divergence with people's perception that warmer conditions and fewer chilling hours are required for plum productivity.

Table 4.12 gives the record of production (MT), area of production (ha) and productivity (MT/ha) of peach in Kotkhai tehsil. As per this data, the area has been increasing and production has demonstrated a non-uniform pattern from the years 2007-2017. The productivity has shown an inconsistent pattern throughout the years. The scatter chart (Figure 4.7) presenting the trend of productivity over the years (2007-2017) indicate towards maximum productivity in year 2009 and minimum in year 2012.

Table 4.12: Peach production (MT), production area (ha) and productivity (MT/ha) in Kotkhai tehsil

Year	Peach production (MT)	Area under peach production (ha)	Productivity (MT/ha)
2007	46	23.34	1.97
2008	56	23.34	2.40

(*Contd.*)

Year	Peach production (MT)	Area under peach production (ha)	Productivity (MT/ha)
2009	72	23.49	3.07
2010	71	29.14	2.44
2011	65.65	31.87	2.06
2012	33.3	31.87	1.04
2013	42.5	30.37	1.40
2014	52.52	31.07	1.69
2015	35	31	1.13
2016	49	31	1.58
2017	37.5	31	1.21

Source: Department of Horticulture (Shimla).

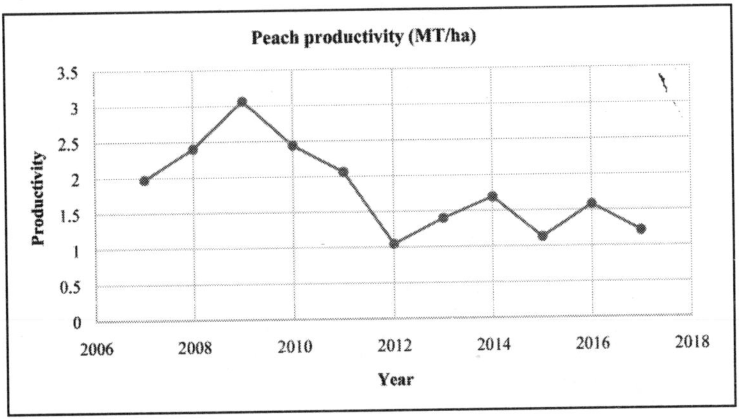

Source: Based on data from Department of Horticulture (Shimla).

Figure 4.7: Scatter chart showing peach productivity (MT/ha).

Table 4.13 reveals the climatic conditions during different growth phase of the fruit for the years 2007-2017 for temperature and rainfall, and years 2010-2017 for snowfall. The high maximum and minimum temperature and less winter rainfall during the initial growth phase has been more for the year of maximum productivity, thus, supporting the perception of the respondents that warmer winters were favourable for stone fruits than pome fruits.

Table 4.13: Productivity and climate conditions during different growth phases of Peach from the year 2007 to year 2017

Year	Productivity (MT/ha)	Pre-flowering, flowering, fruit-set stage (January-March)				Fruit growth and development stage (April-June)			
		Max temp (°C)	Min Temp (°C)	Rainfall (mm)	Snowfall (cm)	Max temp (°C)	Min Temp (°C)	Rainfall (mm)	Snowfall (cm)
2007	1.97	13.6	4.7	92.6	NA	24.4	15.4	20.1	0
2008	2.40	13.9	4.9	0.0	NA	22.6	13.9	0	0
2009	**3.07**	**16.2**	**6.9**	**24.0**	**NA**	**24.8**	**14.7**	**31.0**	**0**
2010	2.44	17.1	7.2	46.7	10.8	26.3	16.2	72.9	0
2011	2.06	14.9	5.3	NA	29.6	23.8	13.8	105.4	0
2012	**1.04**	**13.7**	**4.2**	**63.3**	**7.9**	**25.1**	**14.8**	**4.7**	**0**
2013	1.40	14.6	5.2	120.2	11.6	24.1	15.1	75.0	0
2014	1.69	14.0	4.1	146.3	19.3	24.6	13.9	87.8	0
2015	1.13	14.3	5.3	122.4	15.0	23.9	14.4	29.4	0
2016	1.58	16.1	6.1	92.1	18.2	25.1	15.2	61.2	0
2017	1.21	15.2	5.7	79.4	11.5	24.4	14.8	128.8	0

Source: Based on data provided by IMD and Department of Horticulture (Shimla).

Table 4.14 depicts the record of production (MT), area of production (ha) and productivity (MT/ha) of apricot in Kotkhai tehsil. As is evident from the data, the area and production have shown a non-uniform pattern from the years 2007-2017. The productivity has followed an irregular pattern throughout the years. The scatter chart (Figure 4.8) demonstrating the trend of productivity over the years (2007-2017) indicate towards maximum productivity in the year 2011 and minimum in the year 2010.

Table 4.14: Apricot production (MT), production area (ha) and productivity (MT/ha) in Kotkhai tehsil

Year	Apricot production (MT)	Area under apricot production (ha)	Productivity (MT/ha)
2007	53	57.31	0.92
2008	63	57.7	1.09
2009	63	58.99	1.07
2010	**32**	**63.2**	**0.51**
2011	**109.2**	**64.82**	**1.68**
2012	53.43	64.91	0.82
2013	75	63.91	1.17
2014	87.36	63.2	1.38
2015	75	63.04	1.19
2016	98	63	1.56
2017	75	63	1.19

Source: Department of Horticulture (Shimla).

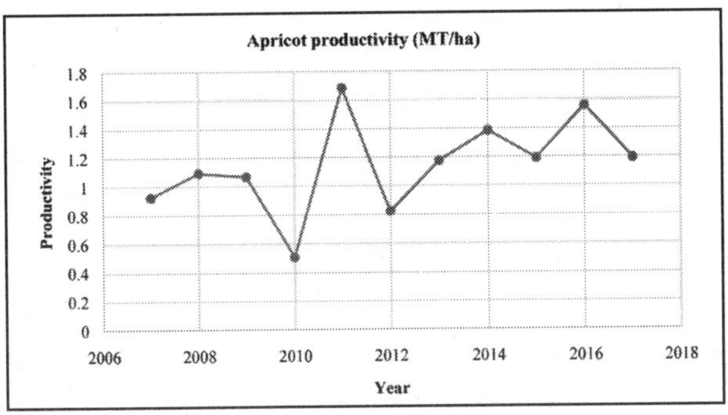

Source: Based on data from Department of Horticulture (Shimla).

Figure 4.8: Scatter chart showing apricot productivity (MT/ha).

Table 4.15: Productivity and climate conditions during different growth phases of apricot from the year 2007 to year 2017

Year	Productivity (MT/ha)	Pre-flowering, flowering stage (January-March)				Fruit-set, fruit growth and development stage (April-June)			
		Max temp (°C)	Min Temp (°C)	Rainfall (mm)	Snowfall (cm)	Max temp (°C)	Min Temp (°C)	Rainfall (mm)	Snowfall (cm)
2007	0.92	13.6	4.7	92.6	NA	24.4	15.4	20.1	0
2008	1.09	13.9	4.9	0.0	NA	22.6	13.9	0	0
2009	1.07	16.2	6.9	24.0	NA	24.8	14.7	31.0	0
2010	**0.51**	**17.1**	**7.2**	**46.7**	**10.8**	**26.3**	**16.2**	**72.9**	**0**
2011	**1.68**	**14.9**	**5.3**	**NA**	**29.6**	**23.8**	**13.8**	**105.4**	**0**
2012	0.82	13.7	4.2	63.3	7.9	25.1	14.8	4.7	0
2013	1.17	14.6	5.2	120.2	11.6	24.1	15.1	75.0	0
2014	1.38	14.0	4.1	146.3	19.3	24.6	13.9	87.8	0
2015	1.19	14.3	5.3	122.4	15.0	23.9	14.4	29.4	0
2016	1.56	16.1	6.1	92.1	18.2	25.1	15.2	61.2	0
2017	1.19	15.2	5.7	79.4	11.5	24.4	14.8	128.8	0

Source: Based on data provided by IMD and Department of Horticulture (Shimla).

Table 4.15 shows the climatic conditions during different growth phase of apricot for the years 2007-2017 for temperature and rainfall, and years 2010-2017 for snowfall. The maximum and minimum temperature during the initial growth phase has been highest for the year of minimum productivity (2010). The highest amount of snowfall has been recorded for the year of maximum productivity (2011). The findings are in divergence with the perception of the respondents that stone fruits productivity is not affected by lesser chilling hours and warmer conditions.

Table 4.16 reveals the record of production (MT), area of production (ha) and productivity (MT/ha) of cherry in Kotkhai tehsil. As is evident from this data, the area has increased till the year 2011, after which there is a decrease in the area. The production has shown a non-uniform pattern from the years 2007-2017. The productivity has shown an erratic pattern throughout the years with maximum productivity in the year 2008 and minimum in the year 2015 as seen in the scatter chart (Figure 4.9).

Table 4.16: Cherry production (MT), production area (ha) and productivity (MT/ha) in Kotkhai tehsil

Year	Cherry production (MT)	Area under cherry production (ha)	Productivity (MT/ha)
2007	265	34.9	7.59
2008	405	38.36	10.56
2009	285	46.13	6.18
2010	263	61.02	4.31
2011	690	76.31	9.04
2012	227.7	76.06	2.99
2013	200	75.16	2.66
2014	414	74.26	5.57
2015	80	74.07	1.08
2016	392	74	5.30
2017	300	74	4.05

Source: Department of Horticulture (Shimla).

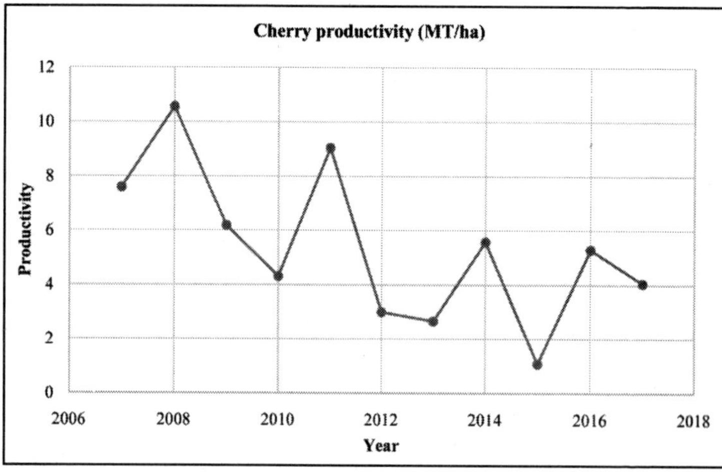

Source: Based on data from Department of Horticulture (Shimla).

Figure 4.9: Scatter chart showing cherry productivity (MT/ha).

Table 4.17 presents the climatic conditions during different growth phase of cherry for the years 2007-2017 for temperature and rainfall, and years 2010-2017 for snowfall. There has been no rainfall during the different growth phases for the year of maximum productivity (2008). The winter rainfall recorded for the year of minimum productivity (2015) is on the higher side. The data did not exhibit any significant correlation with the people's perception.

While the secondary data depicts varying relationship between productivity and climatic conditions for the crops than the perception of the people, it must be kept in mind that the crop production is not solely dependent on climatic conditions, but also on the various steps taken to enhance the production, gradients like altitude and whether the area is sun-facing or lies on the shady side of the mountain. Recently, similar findings were observed by Williams (2021), where he stated that whether the slope is facing sun or not plays a role in creation of the local climate, microclimate, of the slope. Similarly, during the study, it was found that the people perceived that productivity of crop was affected differently at different altitude. The respondents

Table 4.17: Productivity and climate conditions during different growth phases of cherry from the year 2007 to year 2017

Year	Productivity (MT/ha)	Pre-flowering, flowering, fruit-set stage (January-March)				Fruit growth and development stage (April-May)			
		Max temp (°C)	Min Temp (°C)	Rainfall (mm)	Snowfall (cm)	Max temp (°C)	Min Temp (°C)	Rainfall (mm)	Snowfall (cm)
2007	7.59	13.6	4.7	92.6	NA	24.0	14.6	4.5	0
2008	**10.56**	**13.9**	**4.9**	**0.0**	**NA**	**22.6**	**12.9**	**0**	**0**
2009	6.18	16.2	6.9	24.0	NA	23.8	13.3	35.55	0
2010	4.31	17.1	7.2	46.7	10.8	26.6	16.0	57	0
2011	9.04	14.9	5.3	NA	29.6	23.5	12.9	79.5	0
2012	2.99	13.7	4.2	63.3	7.9	23.5	13.3	1.5	0
2013	2.66	14.6	5.2	120.2	11.6	23.9	14.4	41.5	0
2014	5.57	14.0	4.1	146.3	19.3	23.1	12.2	121.1	0
2015	**1.08**	**14.3**	**5.3**	**122.4**	**15.0**	**23.2**	**13.7**	**19.7**	**0**
2016	5.30	16.1	6.1	92.1	18.2	24.8	14.4	53.8	0
2017	4.05	15.2	5.7	79.4	11.5	24.6	14.1	120.6	0

Source: Comparison of data from field, IMD and Department of Horticulture (Shimla).

of Kufar Bag, an upper-middle altitude village, observed that because the slope of the village faced against the sun, the winter were still chillier in comparison to other villages and snow stayed for longer period of time thus benefiting the apple production in the area. While Jashla, though at a higher altitude, lied on the sun facing slope like the other two villages, i.e., Kiari and Bhawana, therefore snow melted away quickly thus leaving less time for water percolation and affecting the snow-fed groundwater level and eventually the productivity of the various fruit crops.

The crop production and its productivity does not solely depend on the climatic parameters, as inferred from the variations observed in productivity with respect of climatic conditions, especially in case of pear. Thus, it can be inferred that there may be other factors responsible for variation in the productivity for those years as stated by Tandzi and Mutengwa (2020) in their study that the yield of harvest crop may be influenced by several factors that can be grouped into three categories, i.e., technological (agricultural practices, managerial decision, etc.), environmental (climatic conditions, soil fertility, topography, water quality, etc.) and biological (diseases, insects, pests, weeds).

5

Impact of Climate Change on the Livelihood

An Emic View

This chapter comprises of the analysis of the impact of climate change on the livelihood by taking into consideration the perception of people about the climate change's impact on different assets of livelihood. It deals with the fourth and fifth objective, to see the impact of climate change on the socio-economic dimension and livelihood of the people of the field area, and to analyse effects of climate change on health condition of people. The impact of climate change on natural assets (land, water and biodiversity), human assets (household member, active labour, education, health, and knowledge and skills), social assets, physical assets, financial assets and horticulture have been taken into account in this chapter.

According to Tuncok (2021), climate change is going to impact four key sectors, i.e., water, agriculture, energy, and environment in Central Asia, which have adverse implications on water security, food security, energy security, human health, etc. The findings of the study suggested that 98 per cent of the respondents were of the opinion that the changing climate affects different assets of livelihood, while 2 per cent did not notice the impact of climate change on their livelihood (Figure 5.1).

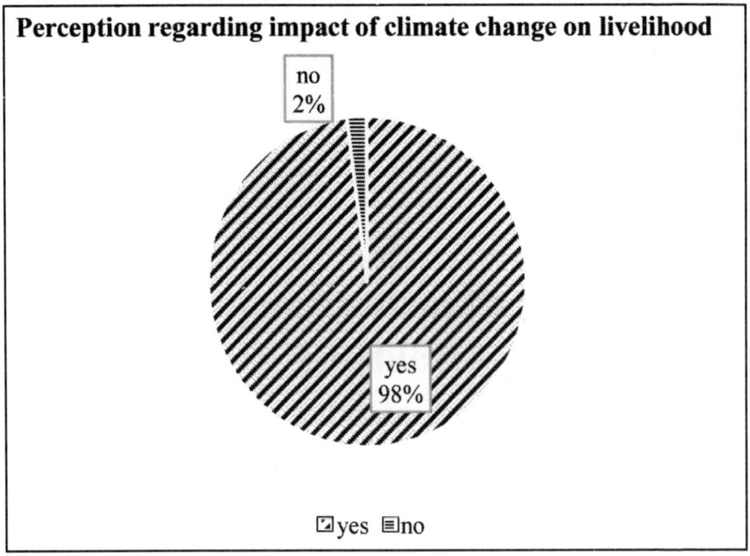

Figure 5.1 Perception regarding impact of climate change on livelihood among the respondents of the study area.

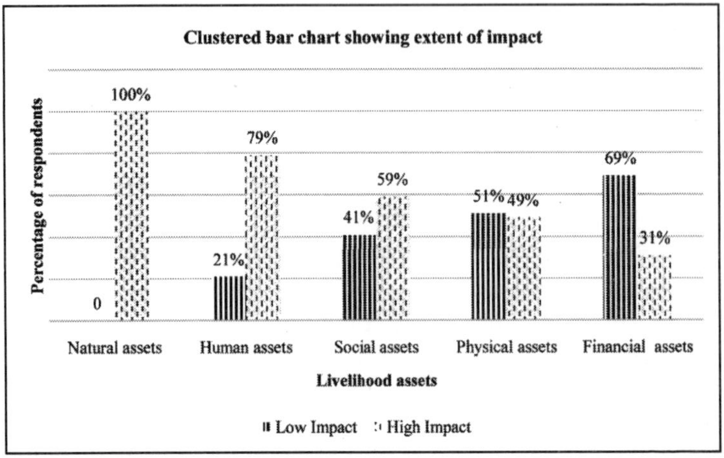

Figure 5.2: Perception regarding extent of impact of climate change on assets of livelihood among the respondents who observed the effect of climate change on livelihood.

Of those who felt the effects of climate change, different extent of impact was observed on the livelihood assets. Figure 5.2

represents the perception of the respondents regarding impact of climate change on the livelihood. All the respondents observed that changing climate have high impact on the natural assets, followed by 79 per cent respondents feeling a high impact on human assets, and 59 per cent observed a high impact on social assets. In case of physical asset, 49 per cent respondents noticed high impact of climate change, while 69 per cent of respondents noticed a low impact on financial assets.

5.1 Impact of climate change on natural assets of livelihood

Natural assets comprise of resources such as soil or land, forests, water and biodiversity. The respondents of the present study observed that the natural resources were the worst affected assets of livelihood, thereby adversely impacted their day to day lives and source of income, which was primarily based on horticulture.

5.1.1 Soil/Land

Due to decline in snowfall and short-lived monsoon, the soil was losing its moisture content. According to some respondent from the lower altitude village Bhawana, lower-middle altitude village Kiari and higher altitude village Jashla, the snow did not stay for longer period as compared to the upper-middle altitude village Kufar Bag, which is a north facing slope. According to the respondent (65-year-old, male, from the low altitude village, Bhawana),

> *Sūkhā paṛane ke kāraṇa zamīna me darāre ā gaī hai.*
>
> (Due to dry conditions, the land has developed cracks.)

The soil was becoming dry as temperature was increasing and rainfall was becoming erratic. The warmer winters did not let the snow stay for longer duration, hence leaving less time for seepage of water. The snow evaporated faster due to increasing temperature in the area. The slope of the hill on which the villages were situated also affected the moisture of the soil. A respondent (71-year-old, male, from the lower-middle altitude village, Kiari), stated that,

> *Mausama garma ho gayā hai. Sardiyon me bhī aba* temperature high *jātā. Hama loga parāle me hai isīliye barpha tikatī hī nahīn aba. Barpha ṭikegī nahīn to* ground water recharge *nahīn hotā. Aise me zamīna kī namī khatm hotī jā rahī hai.*
>
> Climate is warming. Temperature is high even in winters. We are on the south facing slope (*parāle*, facing sun); therefore, snow does not stay anymore. If snow does not stay then the groundwater table will not recharge. This is leading to a decline in moisture content of the soil.

The respondents from the villages on south facing slope (*parālā,* sunny), Kiari and Jashla observed that snow stayed for lesser time now because the temperature was increasing. The respondents from Kufar Bag (*śīlā*, shady as lying on north facing slope) noted that the snow lasted longer in the area which led to moisture retention in the soil. While the respondents of Bhawana noticed that even though the village was on a south-east facing slope, i.e., on both sunny and shady part of the hill, due to negligible snowfall since past few years, the land has dried up. However, all the villages faced the issue of untimely lashes of rainfall which hardly percolated the surface and washed away nutrients along with it as surface run off. Thus, impacting the nutrient composition of the soil of the area under study.

The findings of the present study suggest that the soil is losing its moisture due to increase in temperature and decline in precipitation as well as nutrients were observed to be eroded away by random heavy lashes of rain and incidents like cloud burst. This is in convergence with the study performed by European Environment Agency (EEA, 2015) which highlighted that there are clear signs that soil moisture content is being affected by changes in precipitation patterns and rising temperatures. Also, according to Pareek (2017), soil development procedures and properties related to crop production will be affected by climate in India and will thus influence the livelihoods of millions of people in the country. The people of the study area were facing trouble with the livelihood practice of horticulture due to adverse effect of changing climate on the soil.

5.1.2 Water

Water as a livelihood asset has been the most adversely impacted natural asset in the study area as per the respondent. The respondents perceived that changing climate, like increasing temperature, minimal snowfall, and erratic and untimely rainfall has dried up almost all the natural water sources in the area and there was a significant decline in the groundwater level. This has further affected the health and horticulture practices in the study area. The impact on water is similar to that observed by the United Nations Office for Disaster Risk Reduction (UNISDR)/ SAARC (SAARC) Disaster Management Centre (DMC) (UNISDR/SDMC, 2014) who informed that the unprecedented change in climate have severe effect on the hydrological cycle and water resources, thereby leads to drought, flood, scarcity of drinking water, etc. The impact on water resource have further influenced the food security, health, human settlements and on other related areas in the SAARC region.

The most striking impact on water in the lower and lower-middle altitude villages was the disappearance of *chaśamā or moka* (aquifers) which were the traditional natural source of water in the area. These areas were now dependent on the little water left in the *khaḍa* and *bāoṛī* respectively and lift scheme water supplied by the government through channelling water from water sources in the higher altitudes.

The respondents from Bhawana revealed that scanty rainfall along with increased temperature had dried the natural water sources. Local sources of water had completely depleted since year 2000. The water which was earlier available in their orchards in *bāoṛī* was finished now. One of the respondents (51-year-old, male, from the lower altitude village, Bhawana), recalls,

> Rainwater harvesting *shurū kara dī thī maine jaba merā baṛā betā* 8th *me thā.*
>
> I started rainwater harvesting when my elder son was in 8th class.

The respondent (54-year-old, male, from the lower altitude village, Bhawana) remembered,

> *Chhote chhote nāle ikaṭṭhe hote the* naturally catchment *ke sātha khaḍo me. Inme pānī kī* seepage *hotī thī jo bāda me* everyday use *ke liye kāma ātā thā. Eka khaḍa sa bānaatā thā bāriś ke pānī ke* force *se, itanī bāriśa hotī thī. Vo eka* natural check dam *bānaa jātī thī. Eka eka karake* check dam fill *hotī thī. Pahale ūpara vālā phira bāda me nīche vālā area.* See page *se* moisture *rehatā thā. Bhādo mahīne me zyādā pānī kī vajaha se moka miṭate hote the. Aba* water sources are not getting recharged. *Ūpara se aba* roads *bhī* metal *ho gayī hai jiname pānī kama rachatā hai.*
>
> Small streams collected in catchments in naturally formed large pits. The accumulated water seeped into land which was later used for daily activity purposes. A large pit used to form by the force of rainwater, it used to rain this much before. That would become a natural check dam. The check dams would fill one by one. First in upper areas then in lower areas. See page would retain the moisture. Due to excessive water in the month of August-September, aquifers, *chaśamā/ moka*, were formed. Now the water sources do not get recharged. Besides that, now the roads have become metalled which does not allow much water to percolate.

Similarly in the lower-middle altitude village, Kiari, the respondents observed that over the years, the natural water sources had dried up due to decline in snowfall and rainfall. A respondent (61-year-old, male, from the lower-middle altitude village, Kiari) observed that,

> Height *me* check dams *nahīn rahen.* Bhuilaghat *se* Kanarghati *taka unnīsa kuphara the. Ab do hī reha gaye. Ek* Dhalon *me hai, bākī* ground *bānaa gaye hai. Pahale bāoṛī me pānī* filter *hoke jātā thā, mīṭhā pānī hota thā.* Jungle *me pahale talāb the vo ghata gaye.* Dhalon *kā pānī* seepage *se nīche jākara* Deori *kī bāoṛī me nikalatā.* Kiari *kī bāoṛiyān aba* dry *hai para* Deori *kī nahīn. Kyonki pānī nīche jā rahā hai zamīna me. Basa* Halaibao *me Mātā kī bāoṛī me pānī hai kyonki Devī Devatā ne bachā ke rakhā hai.*

> There are no check dams left in heights. There were nineteen *kuphara* between Bhuilaghat and Kanarghati. Now there are only two. One is in Dhalon, rest have become grounds. Earlier water used to get filtered and reach the *bāoṛī*, the water tasted sweet. There were ponds in the jungle that have reduced. Dhalon water seeps and reached the *bāoṛī* in Deori. The *bāoṛiyān* in Kiari are dry now unlike Deori. Because water level is going down in ground. Only *Mata ki Bāoṛī* in Halaibao has water as the Gods have protected it.*

According to a respondent (84-year-old, male, from the lower-middle altitude village, Kiari),

> Global warming *ne sārā kāma kharāba kara diya hai. Hamare* time *me* snowfall 4-feet *taka bhī hotā thā,* height *para to* 6-feet *taka. Aba* 6-7 inch *hī girtī barapha.* Climate change *se* orbit warm *ho gayā hai. Bāriśe to khatma hī hai. Bāriśa taba hogī jaba thīka* time *para kudarata kā* temperature maintained *rahegā.* Climate change *se saba chaśame* dry *ho gaye hai āsa pāsa. Nāle the hamare, jise* brook *kahate. Saba me pānī nikạlatā thā. Bhraun kī āvāza ātī thī nālo se barsāta me. Aba to* boring machine *se* pump *kara ke pānī nikālate hai. Nālo kī jagaha aba nāliyān hai. Pahale nāle se hī kuhl bānāke le jāte the* gravity *kī* help *se pānī ghara taka. Aba to eka jagaha pānī pahale ikaṭṭhā karnā paṛatā hai, phira vaha se le jāte* individually. *Gītā me bhī likhā hai ke jahan rehāish ho vahān pānī kā* source *hona chāhiye. Pahale mandira me bāoṛī thī jo aba sūkha gaī. Pahale* time *para peṛa hote the jinake patte chauṛe hote the aur vo vātāvaraṇ se namī khinchate the jinase zamīna me namī bānī rehatī thī. Kaniṣka ke peṛa hote the hamare pahale. Vahān tīn jagaha pānī hi pānī thā. Jaba vo kāṭa diye seba lagāne ke liye, taba vahān sūkha paṛa gayā.* Kangi Nala *se ātā hai aba pānī yahan. Aba to ye hāl*

* Bhuilaghat and Kanarghati are the areas that lie above Kiari village. Dhalon is the area where the *kuphara* (check dam) that supplies water to natural water sources lies. Deori is the village that lies below Kiari's altitude. Halaibao is the market area of Kiari. *Mata ki Bāoṛī* is the name of the water source in Haliabao.

> *hai ke* with climate change we will soon be living in Rajasthan state.
>
> Global warming has adversely affected everything. During our time, it used to snow 4-feet as well, in higher areas it used to be 6-feet. Now it only snows 6-7 inches. Climate change is warming the orbit. Rains are almost over. It will rain when the temperature of the nature is maintained. All *chaśame* (aquifers) have dried nearby. We had small streams that we call brooks. All were filled with water. They used to make noise during rainfall. Now we use boring machines to pump water. Streams have been replaced by pipes. Earlier we used to make *kuhala* (water channels) from the streams and take water home with the help of gravity. Now we have to collect water at a place from where it is taken individually. It has been written in the *Gita* also that there should be a water source near the place of habitation. Earlier there was *bāoṛī* (step-well, a water source) in the temple which dried out. There used to be broad leaved trees earlier that took moisture from the environment that maintained moisture in the soil. We had trees of *Kaniṣka* (Alnus). There was abundant water at three places. When we cut them for apple plantation, that area dried out. We receive water from Kangi Nala now. We will soon be living in Rajasthan state with this extent of climate change.

While in comparison to Bhawana and Kiari, there were still natural water sources available in upper-middle altitude village Kufar Bag. Though with time these resources (*chaśame* and *kuphara*) have considerably depleted. The village Kufar Bag was named after multiple *kuphara* present there. Over the time, only 2 *kuphara* are left in the area. Even in the same village water scarcity was different at different height. When the respondents were asked about water scarcity, respondents living at the higher region felt the maximum impact of water scarcity. While the respondents living in the lower region were slightly affected by decrease in water level due to declining snowfall and rainfall.

A respondent (65-year-old, male, from the upper-middle altitude village, Kufar Bag) who lived at the top of the village, gave some interesting insights about how over the years, the decline in snowfall and increase in temperature have led to shortage of water. He stated,

> Lower belt *kī jagahon me to* problem *hai hī, yahan bhī bīcha* jungle *me jo* sources *the sūkha gaye hai. Isa sāla* (2018) lowest snowfall *hua hai* till date. Water sources *kī* conditions worst *hai. Pānī ke pīchhe jhagaṛe hote hai.* Daily use *aur bāgīche ke liye* gutter *kā pānī* use *karate hai yā phira jo* 14-15 *dino me eka bāra* Kangi Nala *ka pānī ātā hai usa para* depend *rahate.* Our elders were scientific thinkers. *Unhone kuphara bānaaī* as check dams. *Isame pānī ikaṭṭhā hota thā jo* water sources *ko* rejuvenate *karatā thā. Ye hī pānī vo bhī* use *karate the jab paśu charāne jāte the.* But *aba* monsoon *me pānī barasatā hī nahīn kī pānī ikaṭṭhā ho sake aura* groundwater table *taka pahunch sake. Ek* Dhalon *kī kuphara hī hai jo chalī huī hai aura* water sources *ko* recharge *karatī. Usakā bhī* level *kama ho rahā, para itanā nahīn jitanā ghātī wālī kā ho gayā hai.*
>
> Areas in lower belt are facing problems, even here the sources (*chaśame*, aquifers) that were in the middle of the jungle have dried out. This year (2018) witnessed the lowest amount of snowfall till date. Conditions of water sources are worst. People fight because of water. We use water from gutter or depend on water supply from Kangi Nala which comes once in 14-15 days for daily activities and orchard uses (Plate 5.1). Our elders were scientific thinkers. They made *kuphara* as check dams. Water got accumulated in it which rejuvenated the water sources. Same water was used by them when they took their animals for grazing. But now it does not rain enough in monsoon so water is not accumulated and does not reach the groundwater table. Only the *kuphara* in Dhalon has sustained itself and recharges water sources. Even its level is decreasing, but not as much as the one in *ghātī* (Plate 5.2).

Source: Field data.

Plate 5.1: Conservation of water from Kangi Nala and rainwater.

Photo description:*The people in the higher part of Kufar Bag store water from rain and water supplied from Kangi Nala in drums for daily use. The natural sources have dried up due to decline in precipitation and increasing temperature, therefore water harvesting is required.*

Source: Field data.

Plate 5.2: *Kuphara* in *ghātī* with receding water level.

Photo description: *This kufar lies in Kanarghātī or ghātī as called by respondents. The photo depicts the decrease in level of water in this kufar. In past, the water covered the complete area. The water level has receded in the recent time due to decline in rainfall and snowfall, especially snowfall. According to the respondents, the snowfall recharges the water source as it percolates better than the rainwater.*

Another respondent (71-year-old, male, from the upper-middle altitude village, Kufar Bag), recalled,

> 1959 *kī bāt hai. Mai* 10 *sāla kā thā. Chaal kī chotī se falls chalate the. Yahan taka āvāza ātī thī jharano kī. Ītanī zyādā bāriś hotī thī Chaitra mahīne me. Binā chhāte ke* school *nahīn jāte the. Pahale 2-2 mahīne kā* cycle *hotā thā* climate *kā. Phira tīna tīna hua. Aba koi hisāba hī nahīn.* Local cycle *khatma ho rahī bāriś kī. Nāle sūkha rahe. Pahale jagaha jagaha moka the. Aba pānī ke* sources *shrink ho gaye hai. Pānī* 2 km *nīche chalā gayā hai. Isake kai* reasons *ho sakate hai. Hamen moharū, bāna ke peṛo kī zarūrata hai.* But *aba vo bhī nahīn laga rahe.*
>
> It was around 1959. I was 10 years old. There were waterfalls from the hill-top of Chaal. One could hear the sound of waterfall till here. It used to rain a lot in the month of *Chaitra* (15th March–14th April). We did not go to school without umbrella. Earlier, the climate cycle was of 2-2 months. Then it became 3-3 months. Now there is no account. Local cycle of rainfall is ending. Streams and waterfalls are drying. Earlier there were aquifers (*chaśamā/moka*) at many places. The water sources have shrunk now. Water level has gone down by 2 km. There can be multiple reasons for this. We need trees of *moharū* (*Quercus dilatate*), and *bāna* (*Quercus leucotrichophora*). But even they are not growing now.

In Jashla, the high-altitude village, the respondents have not faced much issues related to the water. However, a decline has been observed in the quantity of water in the available sources.

Recently, lift water scheme has also started in the area and rainwater harvesting is being done by people living in lower parts of the village Jashla. As per a respondent (37-year-old, male, from the higher altitude village, Jashla),

> *Pānī kī kamī nahīn hai para* mismanagement *hai.* Climate change *ke chalate āne vāle* time *me dikkata ho sakatī hai. Isīliye hamen abhī se dhyāna denā chāhiye.*
>
> There is no water shortage but wate mismanagement. It might be a problem in future due to on-going climate change. Therefore, we must take care from now only.

According to the respondents, the water sources were recharged in a place or two after a good spell of winter snowfall in the year 2019. However, they did not last for long and eventually dried up after few months. Older sources which were still functioning, demonstrated a decline in quantity.

5.1.3 Biodiversity

Climate change have impacted the biodiversity in the area. Many plant and animal species have become rare in the area. The forest has become thinner due to deforestation and forest fires. The new trees being planted cannot cope up with the existing climate conditions resulting in failure of plantation drives and weak, thin trees. New species of plants are seen infiltering the study area.

Table 5.1 and 5.2 shows the changes in the biodiversity as observed by the respondents. Table 5.1 presents the variations in flora in the study area over the time. The plants like *kāfala* (*Myrica esculenta*), blue berry (*Vaccinium cyanococcus*), *jāmuna* (*Syzygium cumini*) and *banaphaśā* (*Viola canescents*), which were present in abundance in earlier, have become rare. *Śṭākā maśarūma* (*Lentinula edodes*) have vanished and wild strawberries (*Fragaria vesca*) have become fewer with time. *Chaiūn* (*Morchella esculenta*) which were found in abundance earlier, appeared in irregular pattern. The dense clusters of *bāna* (*Quercus leucotrichophora*), *moharū* (*Quercus dilatate*), *kharaśū* (*Quercus semicarpifolia*) and *Kaniṣka* (*Alnus nepalensis*) have been thinning and the vigour of these trees have been deteriorating.

Table 5.1: Changes in the biodiversity in the field area (Flora)

Biodiversity			**Earlier**	**Now**
Local name	**Botanical name**	**Common name**		
Kāfala	*Myrica esculenta*	Bayberry	In abundance	Rare
Wild strawberry	*Fragaria vesca*	Woodland or Alpine strawberry	In abundance	Fewer
Blue berry	*Vaccinium cyanococcus*	Blueberries	In abundance	Rare
Jāmuna	*Syzygium cumini*	Black plum	In abundance	Rare
Bāna	*Quercus leucotric-hophora*	Ban oak	Dense clusters with thick and healthy trees	Thinning clusters
Moharū	*Quercus dilatata*	Green oak	Dense clusters with thick and healthy trees	Thinning clusters
Kharaśū	*Quercus semicarpifolia*	Brown oak	Dense clusters with thick and healthy trees	Thinning clusters
Kaniṣka	*Alnus nepalensis*	Alder tree	Dense clusters with thick and healthy trees	Sparse and unable to grow now
Chaiūn	*Morchella esculenta*	Common morel	In abundance	Erratic appearance
Banaphaśa	*Viola canescens*	Himalayan White violet	In abundance	Sparse
Śṭākā maśarūma	*Lentinula edodes*	Shiitake Mushroom	Plenty	Vanished

Source: Field.

Table 5.2 depicts the changes in fauna in the study area with time. The commonly sighted birds like *titara* (*Ortygornis pondicerianus*), *papīhā* (*Hierococcyx varius*), *kauā* (*Corvus culminates*) and *giddha* (*Gypsindicus*) have become rare with time, while *chiṛiyā* (*Passer P. montanus*) has vanished. Animals like *barāgha* (*Panthera pardus*) and *rīkha* (*Ursus thibetanus laniger*) have become rare, and *kakkaṛ* (*Muntiacus muntjac*), *gīdaṛ* (*Canis C. aureus*), *gorala* (*Naemorhedus goral*), *kasturā* (*Moschus leucogaster*), and *kathara* (*Capra aegagrus*) have vanished. On the other hand, *Phīla* (*Gastropoda*) has become common in the recent years.

Table 5.2: Changes in the biodiversity in the field area (Fauna)

Biodiversity			**Earlier**	**Now**
Local name	**Zoological name**	**Common name**		
Titara	*Ortygornis pondicerianus*	Grey francolin	In abundance	Rarely sighted
Papīhā	*Hierococcyx varius*	Common hawk–cuckoo	Commonly heard on rainy days	Rarely heard
Barāgha	*Panthera pardus*	Common leopard	Commonly sighted	Rarely sighted
Rīkha	*Ursus thibetanus laniger*	Black bear	Commonly sighted	Rarely sighted
Kakkaṛ	*Muntiacus muntjak*	Barking deer	Sighted at times	Vanished
Gīdaṛ	*Canis C. aureus*	Jackal	Commonly sighted	Vanished
Gorala	*Naemorhedus goral*	Himalayan goral	Commonly sighted	Vanished
Kasturā	*Moschus leucogaster*	Musk deer	Sighted at times	Vanished
Phīla	*Gastropoda*	Slug	Rare	Frequently sighted
Kauā	*Corvus culminatus*	Jungle crow	Commonly sighted	Rare
Giddha	*Gyps indicus*	Vulture	Commonly sighted	Rare
Chiṛiyā	*Passer P. montanus*	Sparrow	Commonly sighted	Vanished
Kathara	*Capra aegagrus*	Wild goat	Commonly sighted	Vanished

Source: Field data.

The respondents attributed the changes in biodiversity to the changing climate which have affected the flora and fauna of the forests.

A respondent (55-year-old, female, from the upper-middle altitude village, Kufar Bag) noticed that,

> *Aba jungle me vaisī chīzein nahīn milatī jaise pahale thī. Kāphalaaba besvāda hote hai.*
>
> Forests do not have things like before. *Kāphala* (*Myrica esculenta*), are tasteless now.

According to a respondent (81-year-old, male, from the upper-middle altitude village, Kufar Bag),

> *Chiṛiyā eka bhī nahīn dikhatī. Kaue nahīn dikhate. Giddha bhī khatma hī hai. Eka bāra mandir ke pāsa barāgha marā milā thā.*

Not even one *chiṛiyā* (*Passer P. montanus*), can be seen. *Kaue* (*Corvus culminates*), are also not seen. *Giddha* (*Gyps indicus*), have vanished too. Once there was a dead *barāgha* (*Panthera pardus*), near the temple.

A respondent (79-year-old, female, from the lower-middle altitude village, Kiari) recalled,

> *Papīhe kī kitanī āvāza ātī thī pahale. Patā chala jātā thā bāriś hone valī hai. Papīhā bāriśa kā pānī hī pītā hai sirpha. Aba nā bāriśa hotī nā unkī āvāza ātī utanī.*
>
> The voice of *Papīhā* (*Hierococcyx varius*), was common earlier. One would know it was going to rain. It drinks rainwater only. Now, there is neither rainfall nor one can hear them that much.

As per a respondent (57-year-old, female, from the higher altitude village, Jashla),

> *Pahale to kathara, gorala, rīkha, barāgha saba hote the jungle me. Ab nahīn patā kuchha. Chiṛiyā bhī gayāba ho gaī hai.*

> Earlier *kathara (Capra aegagrus), gorala (Naemorhedus goral), rīkha (Ursus thibetanus laniger), barāgha (Panthera pardus),* all were there in the forest. Now we do not know. *Chiṛiyā (Passer P. montanus),* has also vanished.

Another respondent (54-year-old, female, from the upper-middle altitude village, Kufar Bag),

> *Phīlen patā nahīn itanī kyun baḍhā gayī hai. Kuchha sālon se ghāsa bhī badala gayā hai. Patte vālā ghāsa hotā hai aba, badā jaldī baḍhatā hai. Kāṭate hi raho.*
>
> I do not know why *Phīla,Gastropoda* have increased so much. From some years grass is also different. Leafy grass is there, it grows very fast. One just keeps on cutting it.

These narratives indicate that the quality of wild fruits available in the forest have declined with time. People observed a loss of old biodiversity and inception of invasive grass varieties and species like *Phīla* (slug, *Gastropoda*).

Forest fires have raised concern in the study area. Respondents noticed that an increase in the temperature and lack of moisture in soil, made it easier to catch fire in forests. Trees might survive these fires but the bushes were all burnt. A respondent (65-year-old, male, from the upper-middle altitude village, Kufar Bag) observed that,

> Temperature *zyādā hone kī vajaha se* bushes dry *ho jāti jo jala jāti.* Jungle *me peṛa hi zarūrī nahīn* bushes *bhī chāhiye. Bāndara aba* directly *bagīcho me seba para* attack *karate hai kyonki* jungle *me kuchha milatā nahīn khāne ke liye.*
>
> Due to increased temperature the bushes are drying which burns. Only trees are not important in the forest, bushes are also needed. Monkey now directly attacks the apple orchards as there is no food left in the jungle. Earlier they used to eat berries. Now they come to eat apples so we have to lookout for any damage that they may cause to the crop.

Grassland fires have equally been a cause of concern in the lower belts due to scorching heat and lack of moisture. According to a respondent (42-year-old, female, from the lower altitude village, Bhawana),

> TV *par dekhate hai hama* news *me kī kitanī bhayānaka chīza hai* climate change. *Pichhale sāla* (2017), *mere ghara ke pāsa ghāsani me laga gaī thī āga. Vo to logon ne dekha liyā* time *para to nukasān nahīn huā zyādā. Varanā ghara bhī jala jātā. Āga to lagatī hī rehatī hai yahan aura āsa pāsa vāle* area *me.*

> We watch in news on TV how dangerous climate change is. Last year (2017), the grassland around my house caught fire. As people noticed it on time, not much damage was done. Otherwise, the house too would have burnt down. Cases of fire are common in this and nearby areas.

5.2 Impact of climate change on human assets of livelihood

The human assets of livelihood, which are, the household members, active labour, education, health, knowledge and skills have been adversely affected by the changing climate. The respondents identified that changing climate had impacted the household, as members, especially the younger generation, were moving out as the future of horticulture was climate dependent and hence, uncertain. The active labour was being affected both in context of time and workload due to extreme weather conditions and inconsistent pattern of climatic parameters. Due to extreme weather events like excessive heat, erratic lashes of rainfall leading to cloud busts, and hailstorm made it difficult to work in the field. The people have become 100 per cent dependent on the labourers. The labourers, which are the Nepali people, often face problems due to such climatic events as they do not have proper houses to live in, thus making them hesitant to work in areas that are prone to such events. Also, the labour wages have increased over the time and once they have enough money, they go back to Nepal for better education for their children and starting their own work. Overall health of the people has been affected by the changing climate as per the people. Though data pertaining to health effects of climate change were not available at the health centres in the study area, people perceived that the climate change has deteriorated the health of the people. Also, with time, new sets of skills and knowledge has been learnt to combat climate change.

5.2.1 Household members

A trend of out migration was observed among the younger generation in the field area. The finding corroborates with World Meteorological Organization (WMO, 2021) that the climate and weather events have led to significant population movements. The people were moving towards cities for better education and job opportunities. According to a respondent (34-year-old, male, from the lower altitude village, Bhawana),

> *Seba laga nahīn rahā to bāhara jānā hi paṛegā. Seba mausama ke bharose hai aura hama seba ke. Kuchha nā kuchha to karnā paṛegā.*
>
> Apple is not growing, so one will have to move out. Apple is dependent on weather and we are on apple. Something has to be done.

As per another respondent (51-year-old, male, from the lower altitude village, Bhawana),

> *Insāna kī* nature *me lālacha hai. Lālacha kīsi chīza kā bhī ho sakatā*—money, glamour. Foreign *gaye to vāpasa nahīn āte loga. Jaba* better *mila rahā to kyu āenge. Āja kala ke bachche bāhara hī baṛe hue hai to unko itanī* attachment *bhī nahīn zamīna se nā vo kheta me kāma kara sakate itanā.*
>
> Greed is in human nature. Greed could be of anything—money, glamour. Gone to foreign, they do not return. When they are getting better then why will they come back. Children of today have been brought up in cities so they are not attached to the land and they cannot work as much in the orchards.

One of the respondents (77-year-old male, from the lower-middle altitude village, Kiari) told that,

> *Gāon vālo ko apanā aura apane* resources *kā dhyāna khuda hī rakhanā paṛegā kyonki nayī* generation *bāhara jā rahī hai achchhī jīvikā kī talāśa me. Kuchha logon ko lagatā hai ki āne vāle samaya me* apple-based livelihood *nahīn chalegā.*
>
> People of the village must take care of themselves and their resources and think for themselves as the younger generation is moving out for better livelihood opportunities. Some of them think that the future of apple-based livelihood is bleak in near future.

These narratives highlight changes not only at household level, but the social structure of the study area has also been affected. The study area comprised mostly of the elderly people.

5.2.2 Active labour

The respondents were of the opinion that the changing climate has led to increased temperature thereby making work in the fields more distressing. Extreme weather events like heavy rainfall with cloud burst or hailstorm can result into injuries thus affecting the work. Also, with the advent of apple production as a cash crop in the field area, there has been an increase in economy which have made people dependent on luxurious and labour free lifestyle over the time. According to a respondent (51-year-old, male, from the lower altitude village, Bhawana),

> *Hama logon ne kāma karnā chhoṛa diyā hai. Pahale aurate khuda ghāsa lātī thī, ādamī* spray *karate the. Ab* 100 per cent *kāma bānda hai. Aba to Gorakhon se kāma karvāte hai.* Dignity of labour zero *ho gaī hai. Koi kāma nahīn karnā chāhatā. Dekhā dekhī me kāma hi nahīn karate loga, 'vo nahīn kara rahā main bhī nahīn karungā'.* 100 per cent dependency on labour *hai. Aba kala ko kisī vajaha se ye* labour *nahīn milī to bhūkhe rehanā paṛegā.*

> We people have stopped working. Earlier women themselves used to bring grass, men used to spray themselves. Now 100 per cent work has stopped. Now the work is done by the Nepali labourers, *Gorakhen*. Dignity of labour is zero. No one wants to work. They see no body is working so they will also not work, 'he is not working so I will also not work'. 100 per cent dependency is on labour. Now if for some reason this labour is not available then we will starve.

The respondents observed that the labour availability has been affected by the climate change as erratic and extreme weather events causes problems for the labourers who live in *kachchā makāna* or temporary houses. Also, high temperature has affected the working efficiency of the labourers. The workload has also increased along with the time taken to complete a task. According to a respondent (53-year-old, female, from the lower-middle altitude village, Kiari),

> *Kabhī zyādā bāriśa ho gaī to inake ḍeron me pānī bhara jātā. Itanī teza dhūpa me bhī kāma karate hai. Aise me dihāḍī to baṛhanī hī paṛegī. Inahen bhī kamānā hai aura ghara vāpasa jānā hai. Bachcho ko achchhī parvāriśa to saba denā chāhate hai.*

> When it rains heavily, their house gets filled with water. They work under scorching sun. We have to increase the wages in such circumstances. They also have to earn and go back home. Everyone wants to give best nurturing to their children.

A respondent (57-year-old, male, from the upper-middle-altitude village, Kufar Bag) shared his recent experiences saying,

> Spray *ke* time *para bemausamī bāriśa se sāri* spray *barbāda ho jātī, phira dobārā* spray *karanī paṛatī. Kabhī achānaka śāma ke* time *bāriśa ho jātī to Gorakhen kāma roka dete. Aise me dihāḍī laga jātī hai para kāma pūrā nahīn hotā. Hamārā to kharchā baṛa hī rahā hai.*

> Unseasonal rainfall during spray time wastes the spray which has to be done again. Sometimes it rains suddenly during the evenings so the labourers have to stop working. In such scenario daily working attendance is marked as done even though the work is not finished. Our expenses are increasing either way.

5.2.3 Education

Majority of the respondents have sent their children to citics for education. The education of the children who are still present in the study area have not been impacted much by the changing climate directly. The health implication due to climate change have indirectly impacted the education.

5.2.4 Health

The climate change has influenced the physical, mental and social health of the people of the study area as perceived by the respondents. Some respondents perceived an immediate impact of climate change on health. For example, a respondent (71-year-old, male, from the upper-middle altitude village, Kufar Bag), responded,

> Strong of mind stout of heart *nahīn rahā aba ādamī.*
> Man is no longer strong of mind and stout of heart.

While some were of the view that the climate change would have implications on health in near future. Like, a respondent (51-year-old, male, from the lower altitude village, Bhawana) felt that,

> Health incidence as such *nahīn āye abhī taka.* But future *me ho sakate hai* because *pānī khatma hai to* lift *kā pānī* use *karanā paṛa rahā hai jisa se* contamination *ke* chances high *hai.*
>
> Health incidences have not occurred as such. But they may occur in future because as the water sources are drying, water from lift has to be used which have high chances of being contaminated.

The respondents were of the belief that due to changes in the climatic conditions there has been a rise in different types of diseases related to plants and animals. There has been an increase in use of pesticides and chemical fertilizers for promoting the growth of various crops being grown due to various emerging diseases, and unfavourable climate conditions which have led to issues like lack of moisture and nutrients in soil, and excessive moisture or drought like conditions. These chemicals have entered into the nutrient cycle and water cycle which have impacted the health of the people. This perception of people can be supported by the assessment by Lake *et al.* (2012), which states that due to altered conditions for food production, new pathogens may emerge and with use of pesticides contamination will be transferred from environment to food which will have an impact on food safety and its nutritional content.

The respondents believed that the older generation were healthier and stronger in spite of no medical facilities than today's generation as they lived in a clean environment and ate healthy, organic, chemical free food that was grown at home. Some of the respondents are of the opinion that use of insecticides and pesticides contaminate the food and water sources which was a reason for the present increasing rate of cancer cases. Bassil *et al.* (2007), reviewed literature supporting link between pesticide use and cancer and found a positive association between exposure to pesticides and cancer. According to a respondent (82-year-old, male, from the lower-middle altitude village, Kiari),

> *Ādamī kī umra pichhale* 50 *sāla me kāphī kama ho gayī hai aura* climate change *kahīn nā kahīn isakā kāraṇa hai. Āja se* 50 *sāla pahale ke* comparison *me aba bimāriyān kaī guṇā baṛha gaī hai.* Runny nose, watery eyes *jaisī* allergies common *ho gaī hai aba. Bemausami khāne kī vajaha se logon ko kaī prakāra kī dikkate ho rahī hai jaise* immunity *kama honā, thakāvaṭa jaldī honā.* Cancer *jaisī bīmārī baṛa rahī hai.*
>
> Age of human beings have reduced drastically in past fifty years and climate change is the indirect factor leading to it. Over the years number of diseases had

> increased manifold as compared to fifty years back. Allergies leading to runny nose and watery eyes are common now. People are facing ailments like lower immunity, getting tired early, due to unseasonal food being consumed. Disease like cancer is increasing.

According to the respondents, the increasing temperature had made it difficult for people to work in fields during the day time. If proper precautions are not taken, one might suffer from heat stroke, and may suffer from problems of dehydration and headaches which may hinder day to day outdoor activities as people were yet to adapt to these conditions of high temperature. Similar inference has been made by USGCRP (2016), according to which heat exposure can cause exhaustion, and can worsen pre-existing conditions related to cardiovascular and respiratory system.

One of the respondents (42-year-old, female, from the lower altitude village, Bhawana) noted that,

> *Garamiyon me* school *me bachchon ko sira darda aura* dehydration *honā āma ho gayā hai. Bāhara khela kūda vāli* activities *para bhī pharka paṛa rahā hai jisa kī vajaha se* long term *me bachchon kī* mental *aur* physical health *para asara paṛegā. Unake* social skills develop *nahīn honge jo vo khelate samaya* interaction *se sīkhate hai.*

> Headache and dehydration are becoming common in the children in the school during summer. Activities like playing outside are also being affected which might take a toll on physical and mental health of the children. Their social skills will not be developed that are learnt during such interactions.

The scarcity of water due to shortage of annual snowfall and rainfall have affected the health directly by contamination of water used for drinking, daily hygiene and sanitation, and indirectly through their livelihood based on horticulture. Recollecting from the incident of outbreak of jaundice in Shimla in the year 2015-16, a respondent (52-year-old, male, from the lower-middle altitude village, Kiari) stated,

> *Kuchha samaya pahale,* health officers *kī eka* team *ne yahan ke pānī kā jāyaza kiyā thā jisa me patā chalā thā kī yahan kā pānī jānavaro ke pīne lāyaka bhī nahīn hai. Usake bāda se kāphī parivāra, jinahe pānī kī hālata ka anumāna thā, unhone* water purifiers *lagavā liye.*
>
> Some time back a team of health officers examined the water quality of the area and found it unsuitable to be drinking, even by the animals. Since then, most of the families, that are aware of the water quality's condition, have installed water purifiers.

According to a respondent (71-year-old, male, from the upper-middle altitude village, Kufar Bag),

> *Gāon me* gout *ke* cases *baṛa gaye hai ādamiyon me,* especially *garmiyon me. Kāraṇa ye kī pahale gāon kā* climate *ṭhanḍā huā karatā thā aura loga* non-vegetarian diet *liyā karate the jo* protein-rich *thī aura śarāba pīyā karate the. Para aba* temperature *baṛa gayā hai aura saradiyan bhī kama ho gaī hai. Jo loga abhī bhī* same diet *lete hai aur śarāba pīte hai unahe akasara* gout *kī dikkata rahatī hai.*
>
> There has been a spike in cases of gout amongst male members of the village, especially during the summer season. The reason being that the village earlier had a cold climate and the villagers had a protein rich non-vegetarian diet and consumed alcohol. But now there has been an increase in temperature and the duration of winter had also reduced. The people who still have the same diet and take alcohol are usually troubled by gout.

Some of the respondents in Kufar Bag were of the opinion that the recent case of Alzheimer's and memory related problems in the village are the consequence of climate change.

According to a respondent (37-year-old, male, from the high-altitude village, Jashla),

> Forest fire *aur kūṛā jalāne kī vajaha se logon ko* respiratory problems *ho rahī hai.*
>
> (Forest fire and burning of garbage is causing respiratory problem in people.)

The reason for decline in health of children and elderly could be attributed to the quality of air which was deteriorating due to incidence of forest fires. These incidents were becoming frequent in the forests nearby due to increase in temperature and drying patches of land and grass owing to the lack of moisture and increased heat. Also, burning of garbage also leads to respiratory problems like lower vital capacity and allergies as it comprises of toxic waste material, such as tetra packs and containers of chemicals being used in the fields for various diseases related to our crops, along with household waste.

Shortage of water due to decline in precipitation and increased temperature have affected apple production, which is the main source of livelihood for the people in the area. The changing and erratic weather conditions have led to stress and anxiety among the people and has instilled the fear of the unknown among the people whose livelihood is based on weather dependent cash crop economy. Cianconi *et al.* (2020) in their study also highlighted that the climate change can lead to extreme weather, which not only affected the physical health but also the mental health.

Also, due to bleak and uncertain future of apple economy, the younger generation are migrating towards the cities for better livelihood options which have left behind the older people with a sense of loss and isolation. An influx of new cultural traits and ways of life are noticed as the younger people were coming in contact with people from various cultures and were imbibing in their traits, leading to acculturation, mostly in case of interstate marriages. The elderly people have developed a sense of isolation and grief as there was so much changing around them– the temperature, the precipitation, the way different seasons felt, and the culture and their communal bond. According to Straza *et al.* (2018), climate change is responsible for livelihood transition which is leading to cultural changes, and alteration in environment and interactions are changing not only traditional knowledge but also identity.

The scarcity of water in all the villages due to decline in rainfall and no rejuvenation of water sources due to negligible snowfall has created tension among the villagers. The stress generated by water scarcity has led to a decline in the unity and affected the bond between the people of the village, thus affecting the community health. Also, the feeling of oneness and trust is put to test when people outsource water from somebody else's source of water. Earlier all families had their own source of water but due to decline in precipitation, the water sources have dried out. This has led to sharing of water sources which sometimes lead to quarrels over the timing and amount of water being used. One of the respondents (79-year-old, female, from the lower-middle altitude village, Kiari), recalled that,

> *Pichhale sāla* (2018), *do ghara ke logon ke bīcha baṛī bhārī laṛāī ho gaī thī* lift *ke pānī ko leke kyonki pānī thoṛi dera ke liye hī ā rahā thā aur kīsi ko bhī pūrā nahīn paṛa rahā thā.*
>
> Last year (2018), there was a huge fight between the two households over lift scheme water as it came only for a short duration and was not sufficient for anyone.

The respondents explained that such incidents were hampering the peace and smooth function of the village community causing tensions between the people and affecting the communal bond, and increasing the cases of interpersonal grudges. The similar impact of climate change in the form of water insecurity on health have been stated by McMichael and Lindgren (2011). They highlighted that water scarcity hampered domestic and communal hygiene and in extreme situations led to tension and open conflict, with inevitable adverse health consequences.

Figure 5.3 portrays respondent's perception regarding the impact of the climate change on the health of the people. The perusal of all the narratives illustrate how the climate change influenced the physical, mental and social health of the people of area under consideration.

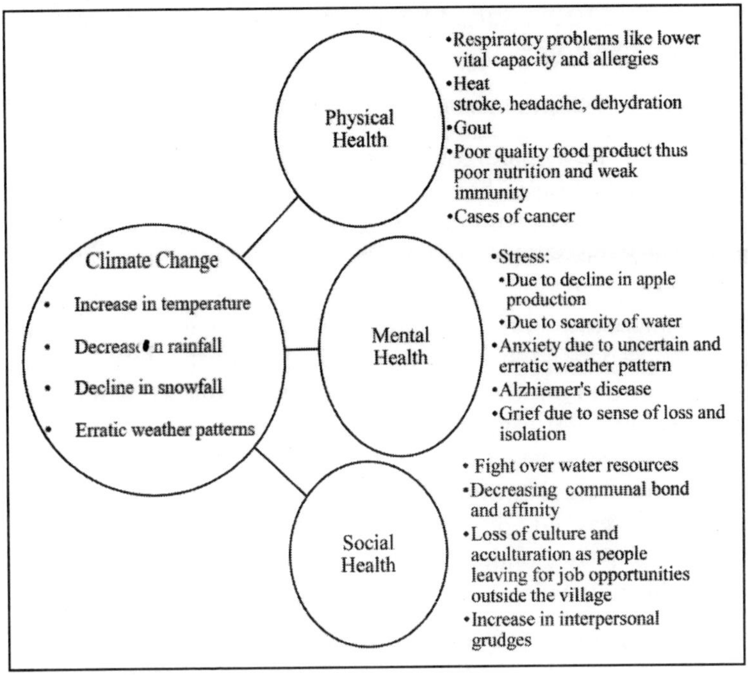

Source: Field data.

Figure 5.3: Climate change's impact on physical, mental and social health of the people.

5.2.5 Knowledge and skills

The respondents felt that new sets of skill and knowledge were to be learned along with keeping up with the traditional knowledge, for instance, traditional knowledge regarding *kuphara*, *kaniṣka*, *bāna* and *moharū* trees. Also, skill like rainwater harvesting and knowledge about various diseases emerging due to climate change is the need of the hour. The respondents in the field area have started using anti-hail net and anti-hail gun to combat the impact of hailstorm. The respondents had knowledge about the new disease emerging in the area and have started using fertilizers and pesticides as per the requirement. The area has seen a widespread use of High-Density Plants (HDP). These comprises of spur and gala varieties of apple. According to the respondent, HDP varieties are better suited for warmer climate as they needed fewer chilling hours.

5.3 Impact of climate change on the social assets of livelihood

Social assets have undergone changes due to climate change. The respondents claim to have the strong *āpasadārī* (brotherhood, communal bond), community or group involvement in activities and strong socio-religious influence. Impact of climate change on these aspects of the lives of the people was also observed by the respondents. Some respondents observed gradual changes in customs associated to death and marriage over the time. During incidences like a death in village, earlier all the women of the village would get together and wash all the clothes, used and unused, in the house of the deceased irrespectively and clean the whole house. Now only women of same clan helped in these chores and washed only those clothes that were used in order to save water. A respondent (54-year-old, female, from the upper-middle altitude village, Kufar Bag) felt that,

> *Kāma bhale hī kama ho gayā ho, auraton kā ghara ke bāhara milana julana bhī kama ho gayā hai. Pahale auraten jaba milatī hotī thī to kāma ke sātha-sātha āsa-pāsa kī khabara kā bhī pātā çhala jātā thā.*
>
> Even though the burden of labour was decreasing, the women were now missing out on the time for bonding with others outside home. Earlier when women used to meet then along with work they used to get to know about news regarding various events.

Also, earlier ceremonies like marriage would take place at home and lasted for five to seven days where all the relatives and villagers used to gather and share their joys and experiences. These days weddings usually took place in a hotel, for two days at the maximum, in order to avoid the trouble of cleaning afterwards as well as due to the busy schedule as everyone was engrossed in their work, be it orchards or any job. Not all people could attend these weddings as someone had to stay back at home for taking care of their land and domestic animals. Old people usually missed out these ceremonies leading to a sense of isolation amongst them.

A respondent (56-year-old, female, of the lower-middle altitude village, Kiari) recalled,

> *Ek bāra me kisī gāonvāle kī laṛakī kī śādi me nahīn jā pāī jisakī vajaha se laṛakī kī mā aura mere riśate kharāba ho gaye kuchha samaya ke liye.*
>
> Once I was not able to attend a marriage of the daughter of a fellow villager which led to sourness in the interpersonal relationship between the mother of the girl and me for some time.

The above narratives demonstrate how changing climate is affecting not only the customs in the study area, but the communal bonds and group activities were being hampered. The women who were normally busy in the household chores were now missing out on the opportunities to gather together and share their experiences and information about various events around them.

The day-to-day activities of women and their social life along with the social lives of elderly have been observed to be affected. Disruption of group activities during times of death and changes in the events like marriages in order to avoid the problem of water scarcity, had a bearing on women and older people in particular. The finding converges with the report of Buechler (2009), according to which climate changes have disrupted informal social networks of the women, elderly, poorest, and women-headed households by affecting sharing of gifts and deployment of labour in Sonora, Mexico.

During the study it was observed that the majority of population in villages comprised of old people, their children would only visit during holidays or during the harvesting of the crop, i.e., the apple season. The respondents mentioned that although the younger generation were still attached to their roots, their villages, the uncertainty of livelihood flourishing in near future due to climate change have forced them to move out, thus creating a social void. There is a chance that the coming generation may forget their roots and communal bonds altogether.

The most prominent social asset during the present research was that of socio-religious influence of the local deities. All the people have strong faith in their deities. According to a respondent (61-year-old, male, from the lower-middle altitude village, Kiari),

> *Sārī bāoṛīyan sūkha gayī, kuphara sūkha gaye, para* Dhalon *kā kuphara aura* Halaibao *me Mātā kī bāoṛī nahīn sūkhī kyonki usakī rakṣā hamare iṣta Devatā kara rahe hai. Logon kī bhī yahī manyatā hai. Isī manyatā ke chalate, jaba bhī mandira kā koi kāma hotā hai, chāhe vo peṛa lagānā ho ya svachchhatā abhiyāna, chhote se baṛā ādamī baṛa chaṛa kara kāma me hātha baṭātā hai. Kyūn? Kyonki ye Devatā kā kāma hai.*
>
> All the *bāoṛī* (water sources) and *kuphara* (check dams) have dried out except for Dhalon's *kuphara* and Halaibao's *Mata ki Bāoṛī* because they are being protected by the *Iṣṭa Devatā* (local deity). All the people have the same belief. It is because of this belief that whenever there is a temple related work, be it planting trees or cleaning drive, every person from small to big participate in the work/activity fully. Why? Because it is deity's work.

Thus, the socio-religious role of local deities has shifted towards nature in order to combat climate change. Religion is the most important aspect of the lives of the people of study area. It played the crucial role in social solidarity and smooth functioning of the society.

5.4 Impact of climate change on physical assets of livelihood

The physical assets have not been affected much in comparison to natural, human and social assets. However, the practice of cattle rearing, which was a common practice in older times have changed drastically. The cattle rearing practice in the area has declined due to shortage of water, and contamination of fodder for the cows. Emergence of new diseases due to ongoing climate change have led to an increase in the use of chemicals in the form of pesticides that has contaminated soil and water. This finding can be supported by Food and Agriculture Organization (FAO, 2016) which concludes that climate change impacts livestock directly through heat stress, increased morbidity and mortality, and indirectly through quality and availability of feed and forages, and animal diseases. According to the respondents, use of chemicals has contaminated the grass used for fodder and water used for drinking. As per a respondent (49-year-old, male, from the lower altitude village, Bhawana),

> *Hamare pāsa* 3 *gāyen hai. Pahale* 4 *thī.* Climate change *ke chalate ghāsa kī* quality *kharāba ho gaī hai. Pānī aur* soil contamination *kī vajaha se gāyon kā chārā aba* poisonous *bāna gayā hai. Isakī vajaha se gāyon kī* health *para affect paṛā hai. Usake peṭa me* gas *bānatī rehatī hai. Hamārī pichhalī gāyen bhī* gas *bānane kī vajaha se mara gaī thī. Dūdha kī* quality *bhī aba kharāba ho gaī hai. Dūdha jaldī kharāba ho jātā hai.*

> We have 3 cows. Before we had 4. Ongoing climate change have deteriorated the quality of grass. Water and soil contamination have made the fodder poisonous. This is affecting the cow's health. Gas keeps on forming in cow's stomach. Our last cow died because of gas in stomach. Quality of milk has declined. Milk goes bad fast.

Also, the outmigration of younger generation in search of better, easier mode of livelihood, have made it difficult to manage cattle. According to a respondent (84-year-old, male, from the lower-middle altitude village, Kiari),

> *Pahale sabake gharon ke bāhara gāyen bāndhī milatī thī. Aba to Gorakhon ne hī rakhī hotī hai, vo bhī isa liye kī dūdha becha kara paise kamā lenge. Gāon me aba buzuraga log hī raha gaye hai. Gāyen kā dhyān nahīn rakha sakate. Kaun pānī ḍho ke lāye, yahān apanā guzārā nahīn hotā. Bachche saba bāhara hi rahate sāla bhara to gāyen nikāla hī dī hai maximum logon ne.*

> In past every house had a cow tied outside. Now only Nepali labourers raise cow that too for earning money by selling milk. Village is mostly comprised of old people only now. They cannot take care of the cow. Who will carry water for them when there is scarcity

> for our own use? Children stay outside the whole year so most of the people have stopped raising cows.

A respondent (55-year-old, female, from the upper-middle altitude village, Kufar Bag) said that,

> *Aba logon ne gāyen nikāla hī dī hai. Logon ke pāsa* time *nahīn. Hamane rakhī hai. Main yahan para* 5-6 *gharo me dūdha bhijavātī hu. Ghī bhī bechatī hu. Para aba pahale jaisā svāda nahīn rahā ghī me. Loga thailī kā dūdha lete hai aba. Para usakā gāyen ke dūdha se kyā mukābalā.*
>
> Now people have stopped raising cows. People do not have time. We have kept them. I supply milk to 5-6 houses. I also sell clarified butter. But the taste is not the same as before. People buy milk packets now. But it is not comparable to cow's milk.

It can be deduced from the above-mentioned narrative that the respondents were of the opinion that the fodder for the cows has been affected by the chemicals in the soil and water. These are contaminated by run-off chemicals and is getting scarce, which has impacted the cow's yield and quality. In recent times most of the people have stopped raising cows due to lack of water and are now dependent on milk in tetra packs, which in comparison to cows' milk have lesser nutritional value. This has led to an increase in expenditure in the form of milk and *ghī* (clarified butter). Earlier the cow manure, used as fertilizer, was available at home. But now people have to buy manure, thus increasing the expenditure on horticulture.

Climate change has impacted not only cattle rearing but also bee-keeping business. The inconsistent weather conditions during the pollination season like sunny days and rain in the evening affected the productivity of the bees, thus impacting the bee industry.

Assets like equipment have also been affected by climate change, for instance, grass cutting machines is being used more frequently due to invasion of weed. Excessive use of these machines often leads to damage to the blades and motors. A respondent (58-year-old, male, from the upper-middle altitude village, Kufar Bag) stated,

> *Maine isa sāla* 2 *machine lī hai ghāsa kāṭane ke liye. Ghāsa baḍha gayā hai aura baṛā jaldī baṛhatā. Eka se kāma nahīn hotā.* Motor *garama ho jātī aur blade bhī ṭūṭa jāte kabhī.*
>
> I bought two machines for cutting grass this year. The grass has increased and grows very fast. One is not sufficient for work. Motor gets hot and blades also break at times.

Other than cattle rearing, bee-keeping and equipment used in orchards, other physical assets like vehicles and houses are also facing the challenge of climate change. The erratic extreme weather event like cloud burst or heavy rainfall have caused damage to various physical assets. Olsson *et al.* (2014) highlighted the damage to physical assets due to climate events is well documented for settlements often built-in risk-prone floodplains and hillsides susceptible to erosion and landslides. In September, 2018, a cloud burst led to damage in Halaibao area of Kiari. The cars in parking area were damaged along with the shops. Similarly, heavy lashes of rainfall clog the pump used for lifting water as the catchment area at lower areas get filled with garbage and dirt that erodes away along with the soil. This affects the supply of water and also contamination of water in field area.

One of the striking impacts was observed on the houses due to increasing temperature. According to the respondent (61-year-old, male from the high-altitude village, Jashla),

> *Pahale garībī thī to saba mila ke rahate the. Pās ke jungle se lakaṛī lāke ghara banāye the. Phira seba āyā. Aba logon ke pāsa paisa ā gayā hai. Sabane RCC ke makāna banā diye hai. Para aba mausama garma hota jā rahā hai. Kabhī bhī ṭhanḍa paṛa jātī kabhī bhī garmī.* In *gharon me ṭhanḍa me zyādā ṭhanḍa lagatī aura garmiyon me zyādā garmī. Eka* time *āyegā logon ko pankhe lagāne paṛege. Kaī logon ne to ghara me* table *fan rakhe hue bhī hai. Ye* tin *ke chhatta zyādā*

> heat reflect *karatī. Pahale* time *me patthara kī chhatta hotī thī. Ina nayī chhatta kī vajaha se garmī aura baṛha gayī hai.*
>
> Earlier people were poor and used to stay together. Houses were made of wood brought from nearby forest. Then apple came. Now people have money. Everyone has made RCC houses. But now the weather is becoming hot. It becomes cold at times and hot at times. One feels colder during winter and hot during summer. There will be a time when people will install fans in their houses. Some people have kept table fans in their house even now. This tin roof reflects more heat. During earlier time, roof was made of slates. These new roofs have increased the heat even more.

According to the narrative, the change in housing pattern have been unfavourable in present milieu of changing climate. The houses made of Reinforced Cement Concrete (RCC) were colder in winter and hotter in summer as compared to the traditional wooden houses. However, with the advent of better road facilities, it became feasible for the people to procure material from the market and construct RCC house. Now equipment like fans are needed in order to keep the temperature down in summer.

5.5 Impact of climate change on financial assets of livelihood

According to Olsson *et al.* (2014) impact of climate change on livelihood could result into loss of profit, unemployment, and poverty. However, not much impact on financial assets were observed as the apple cultivation was still fetching good rates and growing High Density Plants and stone fruits have helped in sustaining the economy in the study area.

With the advent of cash crop economy in the form of horticulture, a boost in the financial status of all the land holders has been observed. Over the time, the market of apple has improved due to introduction of new varieties of apples, and the prices of the crop have gone up as the erratic weather pattern often leads to shortage of good quality apple, thus increasing their value. With the climate change, the productivity may have gone down, but the production has still increased with the help of combating tools and techniques and an increase in the area of production. According to the respondent (50-year-old, male, from the lower altitude belt, Bhawana),

> 50s 60s *me āyā seba yahan* but not as a livelihood option. Common man *ke liye* 90s *se huā ye* cash crop *kī taraha śurū. Pahale* income increase *huī, aba* sustain. *Pahale seba bahuta lagate the to* income *baṛhī. Aba seba zyādā nahīn lagate mausama ke chalate para* market *me dāma baṛha gaye hai.* Balance sheet check *karo to barābara hai* input and output. At times loss *bhī ho jātā hai jaise jaba ole paṛate hai* full harvest *ke* time.

> Apple was introduced in the area in 50s 60s, but not as a livelihood option. Apple became a cash crop for common man in 90s. First there was an increase in income, now it has sustained. Earlier the apple was produced at large scale, so the income increased. Now apple productivity is not that much due to climate conditions but the market rates have gone up. If we check the balance sheet, input and output is equal at times there is loss as well like when there is a hailstorm during the time of full harvest.

The people of lower and lower-middle altitude villages have also resorted to different source of income such as grov'ing crops that are suitable for warmer conditions, owning commercial vehicles and taking up contract related to road development and construction work. Other than that, most of the households had a family member who had a job or was into business other than horticulture.

5.6 Impact of climate change on horticulture

The main source of livelihood for the people of field area is horticultural products, of which apple is the major crop produced, while fruits like pears, cherry, plum, apricot and peach are grown at a very small scale. All the respondents were involved in apple production in the area. Horticulture, being dependent on climatic paramcters like temperature and precipitation, have faced a brunt of climate change. Natural assets of livelihood were found to be affected by changing climate which in turn affected horticulture practices resulting in influencing the human, social, financial and physical assets. All the respondents believed that climate change has affected the fruit production. For instance, the respondents believe that decline in rainfall has led to water scarcity which has an adverse effect on horticulture practice in Kotkhai tehsil which showed consonance with the previous study carried out by Goyal and Surampalli (2018), who observed that changes in precipitation in India will affect water resources. The planning and management in several ways will impact the India's agrarian economy, which relies heavily on the monsoon and availability of water for production.

Productivity of fruit crop has been impacted due to sudden and more intense changes in climate which has affected the flowering, fruit development and yield of the fruit crop. Changes in climatic parameters will affect the fruit productivity. As per a respondent (71-year-old, male, from the upper-middle altitude village, Kufar Bag),

> 1959 *se* 2019 *taka,* 60 *sāla kī* variation *dekhen to patā chalegā kī* temperature *baṛa gayā hai. Barpha ke sātha* sleets *ā jātī, olā aura bāriśa,* pure *barpha nahīn.* Pure *ke liye* temperature *nahīn hai varanā* sleets *kā matalaba nahīn. Aba dekhoge ki Māgha* (15th January-14th February) *me* dry spell *se bīmen kī* sprouting *ho jātī.* Sprouting, flowering *nahīn. Varnā* 15th March *taka hotī thī.* 1989-99 *taka ṭhīka tha mausama, usake bāda kuchha* irrelevant *sā ho gayā hai. Barpha me olā ā rahā.* March *me jo olā ātā thā vo bāda me ā rahā.* March *me barpha paṛa rahī, ṭhanḍa kī vajaha se* frost *bāna jātī phira.* March *me* frost is bad for apple. *Nīche se* heat *ūpara se śīt, kabhī kuchha, kabhī kuchha.* Ultimately pollen *jala jātā pāle se, phira* crop failure. *Isa peṛa me tagaṛā phūla thā aura achchhā* pollinator *bhī. Lekin pahale jaise mausama nahīn hai jahān* step by step cold *se* heat *kī tarapha jā rahe āpa* gradually. *Aba* pollen *kharāba to phūla se pharka nahīn paṛegā. Agara mausama achchhā hai to* setting *achchhī hogī.* Variation *nahīn chāhiye.*

> *Yahān to bāriśa ātī hai aura* pollen *dho ke chalī jātī hai.* 6th *se* 12th April *bāriśa huī* (2018), *sārā* pollen wash *ho gayā. Isase seba kī* overall quality *para bhī pharka paṛa gayā hai.* 30-40 *sāla pahale seba toṛane me zora lagānā paṛatā thā. Aba to aise hī* drop *ho jāte.* Taste *nahīn rahā,* life *nahīn rahī.*
>
> From 1959 to 2019, if we see variation for 60 years, we will find that the temperature has increased. Sleet comes along with snow, hail and rain, no pure snowfall. The temperature is not suitable for pure, or else there is no point of sleet. Now you will observe that due to dry spell the tip sprouts in *Māgha* (15th January-14th February). Sprouting not flowering. Otherwise, it happened till 15th March. Climate was fine till 1989-99, after that it has become irrelevant. It hails with snow. Hail that used to occur in March comes later. Snowfall is occurring in March. Cold leads to frost formation. Frost is bad for apple in March. Heat from below and cold from above, at times something at times something. Ultimately pollen burns due to frost which leads to crop failure. This tree had excellent flower and good pollinator. But we do not have climate like before where you are moving step by step from cold towards heat gradually. Now pollen is bad so flower does not matter. If weather is good, setting will be good. Variation is not required. But here rain comes and washes away the pollen. It rained from 6th to 12th April (2018), all pollen got washed away. This has also affected the overall quality of apple. 30-40 years back it used to take force for apple picking. Now they drop like this only. There is no taste, no life left.

As fruit industry depends on regular and timely cycle of seasons every year, the inconsistent pattern of rainfall and change in temperature and snowfall have affected the crop. The respondents noticed lowering of horticultural produce due to current climatic conditions. The requirement of chilling hours was not being met and untimely rainfall and an overall increase in the temperature was affecting the different stages of crop's growth and development. Similar inference was made by Hazarika (2013) who observed a decline in apple production in Himachal Pradesh due to climate change as the increasing temperature has deteriorated the fruit quality and lack of chilling hours during winter have led to poor flowering, irregularity in flowering duration, pattern of flowering and poor yield in fruit.

According to Datta (2013), melting of ice cap due to climate change has led to reduced chilling effect required for the flowering of many of the horticultural crops in Himalayan regions, therefore, affecting the commercial production of horticulture plants.

5.6.1 Temperature

The temperature has shown an increasing trend in the study area. The respondents observed that storage and shelf life of the produce has been affected as temperature has shown a rising trend. A respondent (59-year-old, male, from the high-altitude village, Jashla) observed,

> *Hamāra seba arāma se Pharavarī taka reha jātā thā. Aba to cold store kī zarūrata paṛa jātī hai.*
>
> Our apple used to stay till February with ease. Now we require cold store.

Increasing temperature have affected flowering and fruit development in the study area. Fruit development and time taken for maturation are being affected by increasing temperature, therefore, there has been a shift in harvesting time, which has been delayed by a week. As per the respondents, for proper colour development, low temperature is required just before harvesting. Due to increase in temperature, apple fails to change into desired colour of produce. Also, high temperature during harvesting season at times lead to sunburn in fruits and delay in leaf fall. According to a respondent (84-year-old, male, from the lower-middle altitude village, Kiari),

> *Āja se* 30-35 *sāla pahale thīka samaya para saba rituen ātī thī.* 10-15 *sāla se* winter *me* temperature *bahuta kama* down *hotā hai. Agara hotā hai to phira kuchha dino me*

> warm *ho jātā hai.* Routine *me* normal temperature *nahīn ho rahā to* domesticated *peṛa para itanā pharka para rahā ke* leaf fall *bhī* late *ho gayā hai.* November *taka harī pattiyān hotī hai kyonki* temperature *down nahīn hota.*
>
> 30-35 years ago, from now all seasons used to arrive on time. For 10-15 years, temperature goes down very little in winter. If it goes down then it goes up in few days. Temperature is not staying normal in routine due to which the domesticated plants are being affected to an extent that leaf fall has also delayed. Green leaves are present till November as temperature does not go down.

Cool temperature is vital in winter as chilling requirement of the fruit must be met. This chilling period during dormant phase has a crucial impact on bud break, flowering, development and therefore on quality of fruit. Insufficient winter chilling period severely affected the crop yields in the field area as lack of chilling hours lead to irregular flowering. This has led to non-uniform crop development which affected the crop size and maturation in the same tree resulting in reduced yield. According to a respondent (64-year-old, male, from the upper-middle altitude village, Kufar Bag),

> *Pahale* flowering *pūre bāgīche me eka sātha hotī thī. Para aba kuchha ṭehaniyon para* flowering, *kuchha para* red tips *nazara ātī hai.* Temperature fluctuate *ho to eka ṭehanī para seba lagatā bākī para nahīn. Aba andāzā nahīn lagā sakate ke pūre peṛa para seba lagegā kī nahīn. Peṛa me hi pharka ā gayā hai, bagīchā to chhoṛo. Pūrā bagīchā sātha me phūlatā thā. Aba* uneven *ho gayā hai* global warming *kī vajaha se kyonki* temperature *me* fluctuation *ho gaī hai. Pahale jaba* full bloom *para* flowering *hotī thī taba* temperature 21-22°C *hotā thā. Aba* 26°C *taka jātā hai.* Lower belt *me to* 30°C *taka jātā hai jo achchhā nahīn hai.*
>
> Earlier flowering used to occur together in the whole orchard. But now one can see flower on some

> branches and red tips on some. In case of fluctuation of temperature, only one branch produces apple while rest of the tree does not (Plate 5.3). Now one cannot estimate if the whole tree will give apple or not. There is a difference in a tree, leave aside the orchard. Whole orchard used to bloom together. But due to fluctuation in temperature because of global warming, it has become uneven. Earlier when the flowering was at full bloom, temperature used to be 21-22°C. Now it is 26°C. The temperature is around 30°C in the lower belts.

The importance of temperature in fruit development can be inferred from these narratives. Apple need sufficient chilling hours in the dormant phase and a constant optimum temperature of around 22°C during flowering and setting period in order to achieve good yield.

Source: Field data.

Plate 5.3: Apple tree during flowering stage.

Photo description: *One may see in this photograph that the tips of the branches in the back are yet to bloom.*

5.6.2 Rainfall

Horticulture is a rain fed practice in all the four areas under consideration. The respondents observed that fruit crop was being affected by changes in rainfall pattern and amount. Apple already uses 100-200 litres of water per tree. A decline in rainfall has led to shortage of water to an extent that the daily requirement of an apple tree has become difficult to meet. The people have shifted to other means of livelihood such as jobs in other sectors. However, the youth who are still interested in their ancestral work are trying to come up with the solution, such as High-Density Crops in lower belts. A decrease in rainfall or delay in rainfall will affect the fruit quality, like lower juice content, smaller size, lower fruit weight and absence of colour. Slight changes in precipitation, its timing and duration, and intensity affects the fruit crop. As per a respondent (52-year-old, male, from the lower-middle altitude village, Kiari),

> Production overall *kama ho gaī hai.* Rainfall *kama ho gaī hai to pānī pūrā nahīn paṛatā.* Apple plant succulent plant *hai.* Fruit *ke alāvā apane liye bhī pānī kī zarūrata hai.* Soil *me pānī hī nahīn. Pahale* 8000 *peṭīyan hotī thī, aba* 1100-1200 *hī hotī.* Plants *sūkha gaye hai. Jaba koi aura lagānā ho to vo chalatā nahīn* because *bāriśon ke dina kama ho gaye hai to* succulent tree *kaise chalegā. Rozi-roṭi para pharka paṛa gayā hai.*
>
> Overall production has declined. Rainfall has reduced so there is water scarcity. Apple plant is a succulent plant. It requires water for itself as well along with fruit. Soil lacks water. Earlier production was 8000 boxes, now 1100-1200 boxes. Plants have dried up. When we plant any, it does not last as rainfall has declined so how will a succulent tree survive. Livelihood has been affected.

Change in rainfall pattern like timing, intensity, onset and withdrawal of monsoon, cloud bursts, and dry spells affect the growth and development of the fruit and also give rise to diseases related to the fruit.

Because of unseasonal rainfall, the fruit growth and development have been influenced. A respondent (51-year-old, male, from the lower altitude village, Bhawana) observed,

> *Garmī me bāriśa ho rahī aura* rainy season *me nahīn. Pahale jaba tīkhī garmī hotī thī buzurga bolate the bahuta baḍhiyā, garmī chāhiye paṛanī. Jaba isa tīkhī garmī ke bāda bāriśa ātī thī taba ekadama khula ke* blossom *hotī thī phasala. Seba ne* winter *me* moisture retain *kiyā hotā thā. Garmī paṛatī thī* March *me. Aba* unseasonal *bāriśh se Jeṭha aur Shaaṛa me* diseases *ho rahe like patajhaṛa. Garmī se* disease *khatma hotī thī.* Vegetative growth *ho rahī* but spur formation *nahīn.* March *me bāriśa ya barpha paṛa rahī. Isase garmī* timely *nahīn mila rahī to* cycle *kharaba ya phira* disease *ho rahe. Pahale* 22-24 December *taka* snowfall *ho jātī thī bachapana me. Aba to* Rajgarh *vāle hāl hai.* Fruit mature *hotā* but no colour. 120 days *to mila gaye hai lekīna rakhā huā tākī* colour *āye.* Flowering time early *hai* but season time same *ya thoṛā* late *hī hai* as colour *ke liye rakha chhoṛate.*
>
> It rains during summer and not during rainy season. Earlier when it used to be scorching hot, elders used to say it is good, it should be hot. When it rained after scorching heat, the crop would fully blossom. Apple would have retained moisture during winter. It used to be hot during March. Now due to unseasonal rainfall in *Jeṭha* (15th May- 14th June) and *Śāḍ* (15th June-14th July) disease like leaf fall is occurring. Heat used to destroy disease. Vegetative growth is occurring but no spur formation. It rains or snows in March. Due to this heat requirement is not being timely met which disrupts the cycle or causing diseases. Earlier it used to snow by 22nd-24th December. Now it is Rajgarh (a lower altitude place) kind of condition. Fruit matures but there is no colour. 120 days have been met

> but the crop has been kept so that colour is achieved. Flowering time is early but harvesting time is same or a little late as the crop is retained for achieving colour.

The narratives indicate towards the importance of water for horticulture, especially apple. The decline in overall rainfall and its inconsistent pattern have hindered the overall development of the crop and its quality. Heat period is also essential for the crop which is being interrupted by the untimely rainfall. For instance, the rainfall during flowering stage drops the temperature that affects the pollination by bees, thus, disturbing the next stages of growth and development.

5.6.3 Snowfall

Snowfall is a prerequisite, especially for apple production. A decline in apple production in lower belts has been recorded as there has been little to no snowfall in that area. At the budding stage apple production requires 1200 to 1500 hours of continuous temperature below 7°C between the months of January and February, but it is not met. According to a respondent (34-year-old, male, from the lower altitude village, Bhawana),

> *Kama* snowfall and *baṛhate* temperatre *ke chalate aba hamane* high density plant *jaise* spur *aur* stone fruits *lagāne śurū kara diye hai. Seba aba ūpara chalā gayā hai aura uskī jagaha logon ne āṛhu aur khuramānī lagāni śurū kara dī hai.*
>
> Due to less snowfall and increased temperature now, we have started growing high density plants like spur and stone fruits. Apple has shifted towards higher belt and people have started growing peach and apricot instead of apple.

Since past few years, hill chill varieties (varieties that require higher chilling hours, 1200-1500 hours) like apple have been switched with low chilling requiring varieties (around 700-1200 hours) like High Density Plants (HDP), peaches, etc. As a decline in precipitation and increasing temperature, the apple in lower belt is shifting towards higher belts and are being replaced by low chilling requiring stone fruits.

Parts of higher altitudes where apple production was not feasible before have now become the perfect grounds for apple production. During the earlier times, the region used to stay under snow cover for a long period of time, exceeding the chilling period needed for plant growth, and thus making it impossible for the apples to grow there. Now, due to lesser snowfall and increase in temperature comparatively to the earlier days, the optimal conditions for apple production are possible. A respondent (37-year-old, male, from the high-altitude village, Jashla) stated that,

> Deorighat, *jo itanī* height *para hai, vahān bhī aba seba lagane laga gaye hai. Pahale to barpha hī rehatī thī vahān. Aba jaldī pighala jātī hai* chilling requirement *ke* according. *Maine bhī lagaye hai peṛa. Thīka* response *de rahe.*
>
> Deorighat, a place at so much height, apple has started growing there. Earlier it was covered with snow. Now it melts earlier according to the chilling requirement. I have also planted trees. They are giving fine response.

The lack of chilling hours for apple was found to be the most prominent reason for the shift in the apple belt from lower to higher altitudes in the present study. But in the higher reaches like that of Deorighat, above Jashla, the melting of snow cover has decreased the chilling hours to the required amount, thus, facilitating apple production at that height. Thus, showing variation in impacts of climate change at different altitudes. Therefore, it can be concluded that changing climatic parameters have shifted the apple production from lower to higher altitude areas, thus opening the opportunity of apple cultivation at higher reaches.

All the factors like temperature, rainfall and snowfall need to be ideal for a good crop. For instance, apple flowering has been irregular over the years due to inadequate chilling during winter resulting in the shift of apple belt to higher altitude. Also, minimal snowfall and low rainfall leads to scarcity of water, and as the temperature is increasing, higher temperature would require more water for good performance of the crops.

A shift in apple belt from lower to higher altitude has been seen which indicates at the shift in ecosystem boundary. The work of Xu *et al.* (2009) also proposed that cascading effects of rising temperatures and loss of ice and snow in the Himalayan region were distressing water availability, biodiversity, and were leading to ecosystem boundary shifts like high-elevation ecosystem changes and monsoonal shifts.

5.6.4 Extreme weather events

Climate change has resulted in uncertain weather conditions. Sudden rainfall during the setting period or no snowfall during chilling period has been a problem. A shift in seasons has been observed which makes it difficult for people and crop to adapt. According to people, when there is proper rainfall and chilling period requirement is achieved due to snowfall and low temperature during winter, the apple production is satisfactory. But in erratic weather conditions, there is a decline in apple production. So, depending on the amount of snowfall, rainfall and temperature, apple production may be good or poor, which has a direct impact on the livelihood.

The uncertain weather also has an adverse effect on the orchards in the study area, as perceived by Hewitt and Mehta (2012) in their study where climate change events like flash floods have severely damaged orchards, roads, and stream embanakments in the Himalayas. For instance, in a cloud burst in September, 2018 in Kiari area, the trees along the course of water flowing from Balidhar-Bhuilaghat were damaged in the process.

In June-July, 2020, due to sudden hailstorm, almost 50 per cent of apple crop was damaged (Plate 5.4 and 5.5), which resulted in lower price in market and lesser crop for sale. According to the people, while there were good winter conditions initially and an estimate of 5000-6000 boxes of apple was there during season, due to sudden hailstorm, the destruction was to such an extent that only 2000-3000 boxes of apples could make it to the market. Due to unusual random and frequent hailstorms in summer months, all the fruits being produced in the area (pome and stone fruits) are getting heavily damaged.

In convergence to the present study, Vidaurre de la Riva *et al.* (2013) also illustrated that events like drought, frost, hailstorms in the Charazani of Andean region of Bolivia, along with landslides and increasing cases of fire, were the main reasons for agricultural loss. They further documented that the climate change had a negative impact as the lives and livelihoods of the people depended on local resources.

Source: Field data.

Plate 5.4: Damaged apple crop.

Photo description: *The hailstorm during crop's growth phase damage the crop. The marks on the fruit leads to lowering of market value.*

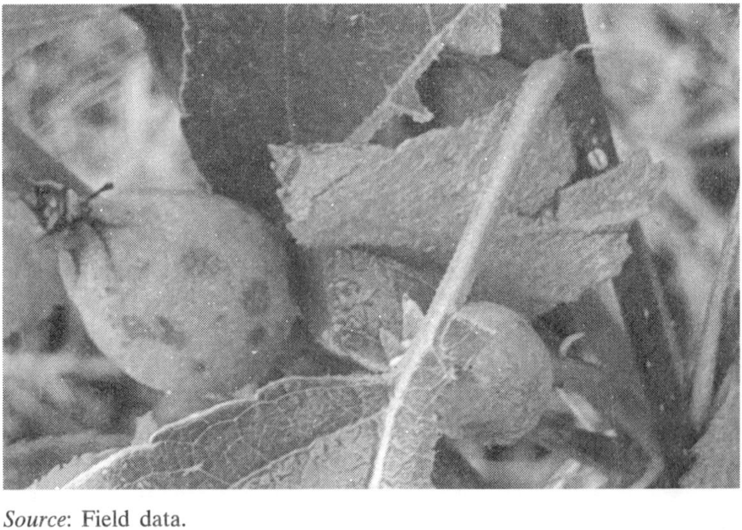

Source: Field data.

Plate 5.5: Damaged leaves due to hailstorm.

Photo description: *The leaves were cut due to high impact of hailstorm which otherwise would protect the fruit from direct intense sunlight. Thus, affecting the colour and size of the crop in later stages.*

Due to increase in the temperature and delay in winter season, there has been incidence of snowfall during flowering period, thereby resulting in severe frost injury. Due to snowfall in April, 2021, which is the flowering period, development of the crop was hampered which delayed the harvesting time by at least 10 days in the entire Kotkhai belt. Also, untimely snowfall caused extra damage in the areas that had installed anti-hail nets to combat issue of hailstorm. The damage can be seen in the photo (Plate 5.6).

Source: Field data.

Plate 5.6: Uprooted tree due to accumulation of snow on the anti-hail nets.

Photo description: *The orchards are covered by the anti-hail nets by the mid of March in order to protect the crop from hail damage. Untimely snowfall in April, 2021 led to uprooting of* *tree as it got collected in the anti-hail net which weighed a lot thus destroying the plant.*

5.6.5 Increase in use of chemicals

Campbell *et al.* (2016), postulated that climate change will have predominant impacts on crop, livestock and fisheries production, and will intensify the prevalence of crop pests. The changing climate has led to change in epidemiology of pests and disease-causing agents. Excess of untimely rainfall and increasing temperature have led to different diseases or pest/insect problems. This has led to increase in the use of insecticides and pesticides, for instance, use of Captan, Dodine and Carbendazim for disease like Scab, and Chlorpyriphos (Pest: Woolly Apple Aphid) and Spiromesifen (Pest: Mites). Due to climate change (less rainfall, higher temperature, inconsistent weather), people are using chemicals to sustain apple production and therefore increased expenditure. Also, in this way climate change is not only impacting livelihood, but also may have negative effect on human health as well as soil productivity and organisms living in it.

According to a respondent (62-year-old, male, from the upper-middle altitude village, Kufar Bag),

> *Pahale* input *kama thā,* nature *sātha detā thā. Phira* chemicals use *karane* start *kiye.* Starting *me achchhā* result *thā* but *aba* output *kama ho rahā. Aba zyādā garmī se* canker *laga rahā, aura gīle se* scab. Lower height *me to* stem borer *bhī laga rahā sūkhe kī vajaha se.* Drought *kī vajaha se* dropping *ho rahī aur pattā pīlā paṛa rahā. Pahale gobara* use *karate the khāda ke liye. Aba* chemical *hī karate. Gobara bhī bāhara se mangvānā paṛatā, gāyen hai nahīn. Aba to hara* 15 *din me* spray *karani paṛatī. Kabhī kabhī to* double *karanī paṛatī bemausami bāriśon kī vajaha se. Kharchā baṛha gayā hai.*
>
> Input was less before, nature use to help. Then we started using chemicals. Result was good initially but now output has lowered. Now due to more heat issue of canker is arising in trees and issue of scab due to excessive moisture. In lower height, issue of stem borer

> is arising due to dry conditions. Dropping is occurring due to drought and leaf is turning yellow. Earlier we used cow dung as manure. Now we use chemical only. Cow dung has to be brought from outside, we do not have cows. Now we have to spray after every 15 days. At time we have to spray twice due to unseasonal rainfall. Expenditure has increased.

Due to change in climate, various issues like canker, a fungal disease (Plate 5.7) has been observed in the field area, which has increased the frequency of spraying insecticides and pesticides.

Source: Field data.

Plate 5.7: Tree with canker, a fungal disease.

Photo description: *The light part of the trees is the part affected by canker. The light colour is of the green coloured fungicide applied to prevent the tree from drying further.*

6

Adaptation Strategies and Government's Efforts to Combat Climate Change

The chapter discusses the adaptation strategies acquired by the people to overcome the climate change and examines the government efforts being taken to combat and mitigate climate change. It comprises of the sixth and seventh objective respectively, which are, to study the adaptations, if any, made by the people due to climate change, and to examine in field area, policies, if any, formed by the government to combat climate change in the given area.

6.1 Adaptation strategies acquired by the people to overcome the impact of climate change

> *"It is not the strongest or the most intelligent who will survive but those who can best manage change".*
>
> *–Charles Darwin*

Though the adaptation process that Darwin referred to was biological adaptation, cultural aspect of the adaptation is equally important for humans as they are the social beings. Culture was defined by Tylor (1871), as "Culture….is that complex whole which includes knowledge, belief, art, law, morals, custom, and any other capabilities and habits acquired by man as a member of society." Adaptation strategies acquired and learnt by the people of the field area are thus a part of the cultural adaptation to the changing climate which has affected the various aspects of their livelihood.

Adaptation measures varied depending on the extent of the impact of climate change on the various assets of livelihood and the requirement of the people of the field area. There is no 'one size fits all' solution to the effects of climate change on the livelihood of the people of field area. The findings of the present research suggest that the impact of climate change is mostly experienced on water and horticulture practices, which is different in extent at different altitude villages. Therefore, adaptation strategies related to the major issues faced by the respondents of different villages have been acquired in accordance to combat climate change impact.

6.1.1 Adaptation strategies acquired by the people for the lower altitude village Bhawana

The lower altitude village of Bhawana has observed the prominent change in the climatic parameters in the past years. The respondents of Bhawana experienced that climate change in the form of increasing temperature, minimal to no snowfall, and declining and erratic rainfall had the major bearing on their water sources and crop production. In order to combat the adverse effects of the changing climatic parameters, the respondents have incorporated various changes and approaches to adapt to their new setting.

The respondents practice rainwater harvesting to collect water for daily use. Tanks are being built in order to collect rainwater. Water supplied by government sources is stored and used as per need. According to a respondent (48-year-old, female),

> *Kāphi* time *se pānī kī dikkata hai. Aba to ādata ho gaī hai kama pānī me sārā kāma karane kī. Kuchha sāla pahale gāyen nikālanī paṛī. Pānī pūrā nahīn paṛa rahā thā.* Gumma *se dūdha le āte hai ḍabbe vāla. Pānī ke liye* drum *bhara lete hai jo kuchha bāriśa ho usse, kheta ke kāma ā jātā hai.* Government *ne* schemes *shurū kī hai.* Lift *kā pānī ā jātā hai khaḍa se. Usme guzārā ho hī rahā hai abhī to.*

> Water problem has been there for some time. We have become habitual now of doing all work with less water. Some years ago, we had to give away the cow. Water requirement was not being met. We get carton milk from Gumma. We fill up drums for water from rain, it comes in use for the field. Government has started schemes. We are able to survive on it for now.

The people of the village are aware of water scarcity and make a conscious effort at individual level to not waste water. With the development of roads and easy access to market nearby, in Gumma, it has become easier to arrange milk and related products.

The respondents have shifted to stone fruits such as plum, apricot and peaches, and High-Density Plant (HDP) like spur and gala variety of apple from the traditional apple variety. The respondents who had flat land available have started growing HDP (Plate 6.1). High-Density Plant (HDP) do not need long chilling hours as compared to the earlier variety of apples being grown in the area. Also, the roots of the HDP are shallow, so they absorb water from the surface and do not require deeper percolation of water. Thus, they are suitable for the declining rainfall and increased temperature in the area. The respondents are using mulching sheets and mats in order for better absorption of water. A respondent (35-year-old, male) informed that,

> Mulching mats better *hai* mulching sheets *se.* Mats *me* pore *hote hai jisase bāriśa kā pānī* soak *bhī ho jātā hai* and *thoṛī bahuta* evaporation *bhī ho jātī hai.* Mulching sheets *me* pore *nahīn hote para vo sastī hai.*

> Mulching mats are better than mulching sheets. Mats have pores through which rainwater get soak as well as a little evaporation also occurs. Mulching sheets do not have pores but they are cheaper.

Source: Field data.

Plate 6.1: High density plants.

Photo description: *High Density Plants are a dwarf variety and does not cover huge space in comparison to the traditional varieties. Therefore, there is less distance between the two plants. These plants need support, which is being provided by the wires tied to the poles on the extreme ends and the trees, as they have shallow roots. supported Mulching mats are being used by the respondent for better percolation of water.*

The respondents felt that High Density Plants (HDP) required only initial investment. As they are a dwarf variety, not much labour is required in the later stages. The respondent (50-year-old, male) who initiated the plantation of HDP in the village said that,

> Scientist High Density Plants *ko* need of the day *bola rahe hai. Dekhate hai agara sahi hai ye. Maine* 900 saplings *lagayen hai. Iname se* 300 foreign *hai aura bākī apnā* rootstock *hai. Inakī* water requirement *kama hai. Uparale pānī se kāma chala jātā hai.* Shallow roots *hai to* drip irrigation *use kara sakate. Maine* sprinkler *rakhe hue hai. Jaba bahuta* requirement *hotī hai, lagabhaga*

> 15-20 *din bād, taba* use *karate hai. Abhī to bahuta kharchā ā gayā hai.* Initial investment *zyādā hai.* Land prepare *karo,* labour *kī* cost, saplings *kī* cost. 9-10 lakh *ke āsa pāsa kharchā ho gayā hai kyonki* land barren *hai to* machine *se pahale* tilling *karvānī parī aura ye* saplings local *se mehangī hai.*
>
> Scientists are calling High Density Plants the need of the day. Let us see if it is right. I have planted 900 saplings. Out of these 300 are foreign and rest is own rootstock. Their water requirement is less. Surface water is sufficient. They have shallow roots so one can use drip irrigation. I have kept sprinklers. When there is need, after around 15-20 days, we use them then. Expenditure has been a lot for now. Initial investment is more. Prepare the land, labour cost, sapling cost. The expenditure has been around 9-10 lakh as the land is barren so tilling had to be done with machines and these saplings are costlier than the local ones.

The people of the village noticed that with changes in climate their main source of income, that is, horticulture is being impacted. New array of diseases and pests are emerging, therefore changes in the horticulture practices is required. The people are becoming more scientific in their approach. But more efforts are to be put in by the people to combat climate change. People who cannot invest much in HDP, feel that apple will not last long, therefore they are shifting towards stone fruits for commercial purpose.

During the research, it has been observed that all the households have secondary source of income. The people are either working in cities or have taken up construction or builder job. People have bought land at places that are still not impacted as much as Bhawana due to climate change. A respondent (65-year-old, male) stated that,

> *Bagīchā liye hai* Ghoond *me. Vo śīlā* area *hai aura* height *para bhī hai to vahān seba achchhā laga jāyegā.*
>
> I have bought orchard in Ghoond. It is a shady area (north facing slope) and is at a higher altitude so apple will grow well there.

The respondents felt that adoption of new techniques and other modes for generating income is the need of the hour in lower altitudes, where the traditional apple cultivation has failed due to changing climate.

6.1.2 Adaptation strategies acquired by the people for the lower-middle altitude village Kiari

The respondents of the Kiari village recognised that water scarcity is the major issue in the area. Therefore, rainwater harvesting is being done at household level. The people collect rainwater through gutters that have been installed on the roof (Plate 6.2). The water is used for orchards purpose as well as for household use.

Source: Field data.

Plate 6.2: Rooftop gutter.

Photo description: *The rainwater gets collected in the gutter at the end of the roof, which then enters the pipe connected to the water collecting tank.*

Conservation of water at individual level is being done. Endorsing the same, a respondent (74-year-old, male) expressed,

> *Logon ko kama pānī me guzarā karane kī ādata dālanī paṛegī. Āne vāle time me pānī kī problem aur baṛha jāyegī. Main kama se kama pānī use karane kī kośiśa karatā hun. Loga hai kī roza nahānā hai. Are kaunsā vo Gorakhon ke sātha kheta me kāma kara rahe. Main bird bath letā hu. 2-3 dina me nahā liya ḍhang se.*
>
> People will have to get used to surviving in less water. Water problem will increase more in coming time. I try to use minimum amount of water. People have to take bath every day. It is not like they are working with *Gorakhen* (Nepali labour) in the field. I take bird bath (dry bath). One may bathe properly in 2-3 days.

Change in mode of transportation of apples has been observed. While earlier trucks were hired to transport apples to the market, due to decline in apple production people use vehicles like utilities. This has helped them in saving money of transport and also given employment opportunity at local level. Most of the people now own commercial vehicles as source of income in the village. Other than commercial vehicles, majority of the younger generation are found to be working in cities for more reliable source of livelihood.

The respondents realize a need to switch to HDP varieties of apple and stone fruits as they survived in warmer temperature and needed relatively less water than the earlier verities of apple.

According to a respondent (34-year-old, male),

> Young blood *jaba ghara baiṭhegā,* source of income *eka hī hogā jo* nature dependent *hai, to kuchha nā kuchha to karanā paṛegā. Āja* India *me* 35-40 varieties *ā gaī hai* apple *kī jise kama pānī chāhiye.* Apple *se haṭa kara bhī* alternative *ḍhūnḍhanā paṛegā.* Land *kī bāta kare to āne vāle* time *me yahan* stone fruit *hi chalegā.*
>
> When the young blood will have to sit at home and source of income will only be one, which is dependent on nature, then something will have to be done. Today

> there are 35-40 varieties of apple in India that require less water. We need to look for alternatives other than apple. If we talk about land, in coming time only stone fruits will survive here.

The respondents are also trying their hand in organic farming in order to conserve the soil for the next generation. The narratives revealed that at the adaptations being made at household and horticultural level. The educated youth who has returned to villages during the pandemic in the year 2020 and are aware of the impact of climate change and various scientific ways to combat it. They have brought about awareness regarding conservation of the environment and sustainable use of resources at the household level.

6.1.3 Adaptation strategies acquired by the people for the upper-middle altitude village Kufar Bag

Water scarcity and hailstorms were the main problems in the village of upper-middle altitude, Kufar Bag. The people were practicing rainwater harvesting and rooftop gutter system was being used to harvest rainwater. The respondents believed that the traditional apple trees can only survive in the micro-climate of the village. High Density Plants (HDPs) need certain initial requirements, like gentle slope, easy access to road and drip irrigation facilities, which could not be met with in Kufar Bag. Traditional trees were still working so there was no need either. However, plantation of forest trees was required to combat climate change.

According to a respondent (65-year-old, male),

> Individual level *para kāma karanā paṛegā.* Afforestation *karanī paṛegī. Hama* gutter *kā pānī* use *kara rahe aba ghara aura bagīche ke kāma ke liye.* Kangi Nala *kā pānī* 15 *dina me eka bāra ātā hai. Bākī hama ālū bīj rahe hai. Usase* soil fertile *hotī hai.*
>
> We will have to work at individual level. We will have to do afforestation. We are using gutter water for household and orchard work. Water from Kangi Nala

> comes once in 15 days. Rest we are growing potatoes. They make the soil fertile.

According to a respondent (63-year-old, female),

> *Kisane sochā thā ke yahan bhī pānī kī kamī ho jāyegī. Nīche kī tarapha nahīn hai tāngī. Yahān ūpara vāle* area *me hai. Sārā pānī nīche chalā gayā hai.* Gutter *kā pānī* use *karanā paṛatā hai.* Scheme *vāle pānī kā koi bharosā nahīn.* Gutter *vāle pānī me* bleaching powder *dāl dete hai tākī ghara ke kāmo me ā jāye kām. Bākī pānī bachāne ke liye kuchha nā kuchha kara lete. Jaise* machine *kā pānī jo ātā hai kapaṛe dho kara usako* flush *ke liye* use *kara lete hai. Chāvala dāla dhoke jo pānī ātā usko sīnchayī ke liye rakha lete hai. Chāya pattī bhī achchhī hotī hai zamīna ke liye. Merā pudīnā sūkha gayā thā pīchhe jaise. Aba is pānī se thīka ho gayā hai.*

> Who would have thought that there will be water shortage here. Lower parts (of Kuafr Bag) do not face shortage. Here in upper area, there is shortage. All water has gone down (water table). We have to use gutter water. Scheme water has no reliability. We put bleaching powder in gutter water so that it may come in use for household work. Rest we do something to save water. Like using machine water that comes after washing clothes for flushing. The water used for washing rice and lentils is used for irrigation. Tea leaves are also good for soil. My mint plant dried out some time ago. Now it is alright because of this water (tea leave water).

Anti-hail guns have been installed in the surrounding areas. The respondents observed an increase in frequency of hailstorm after the installing of these anti-hail guns, first in 2011 in Bhuilaghat and second in 2015-16 in Baghi village. The respondents felt that due to lack of knowledge about the clouds that caused hailstorm, the people dispersed every cloud in fear of hailstorm. This has led to adverse impact on the expected rainfall. As per a respondent (67-year-old, male),

> *Bādala dikhā nahīn ke eka tarapha se* Bhuilaghat *vāle* gun *chalā dete aura dūsarī tarapha se* Baghi *vāle. Vahān se bādala chhanta jātā aura yahān ā ke barasatā.* 2010 *me* bumper crop *hui thī yahan. Phira* 2011 *me* Bhuilaghat *me* anti-hailgun *lagī.* 2012 *me bahuta olā āyā.* Radar *laga huā hai nahīn, kabhī bhī chalā lete hai* gun. *Rāta ko bhī dhuma dhuma kī āvāza ā rahī hotī hai.*
>
> As soon as cloud is seen, people in Bhuilaghat fire the gun on one side and the people of Baghi on the other side. The cloud disperses from there and it come here and rains. We had bumper crop in 2010 here. Then in 2011, anti-hail gun was installed in Bhuilaghat. In 2012 huge amount of hail came. There is no radar in the gun, they fire it at any time. Even at night one can hear *dhuma dhuma* (loud noise that gun makes when it is fired) sound.

One may conclude from the narratives that knowledge regarding the various new techniques and adaptation measures is required for them to be successful. For instance, anti-hail gun installation alone is not sufficient, one must know the proper use of these guns. The people did not know the difference between the normal cloud and the cloud that may result in hailstorm. Also, there were no radars present in the gun which would have helped in differentiating different clouds. Similarly, people of the village, Kufar Bag were aware about the prerequisites for High Density Plant cultivation and owing to the inability to meet those requirements, they did not practice the HDP cultivation.

It was noted during the field visit that the respondents did not face much problem due to hailstorm before, however, after the instalment of the anti-hail guns in nearby areas, frequency of hailstorm increased. The hailstorm of the year 2020, caused so much damage to the crop that some of the people installed anti-hail nets the next year, 2021 (Plate 6.3 and 6.4). The respondents were using new and effective chemicals being introduced in the market for spray purpose to combat the emerging diseases. Due to invasion of new species of plants in the orchards, it was becoming difficult for the people to manually clear the orchards frequently.

The people have started using grass cutting machines for this purpose as it is less laborious.

Source: Field data.

Plate 6.3: Village Kufar Bag before 2021.

Photo description: *The villagers observed an increase in frequency of hailstorm since 2010 but did not face a very intense hailstorm; therefore, they did not use anti-hail net.*

Source: Field data.

Plate 6.4: Village Kufar Bag in 2021.

Photo description: *After the hailstorm in 2020, almost half of the crop was damaged. As the younger generation were there due to covid restriction, they talked the elders into installing the anti-hail nets like the surrounding villages. The people who were affected the most and also, could afford the expense of anti-hail net, installed these nets. Rest of the villagers are waiting to see the effectiveness of these nets in the area.*

6.1.4 Adaptation strategies acquired by the people for the high-altitude village Jashla

The following adaptive strategies were acquired by the people of the higher-altitude village, Jashla, in response to the ongoing climate change. An anti-hail gun was installed in Deorighat in 2011 in order to combat frequent hailstorm in the area (Plate 6.5). They have now installed anti-hail nets too as the anti-hail gun did not have the expected results.

Source: Field data.

Plate 6.5: Anti-hail gun in Deorighat.

Photo description: *The anti-hail gun was installed in Deorighat in 2011in order to tackle the issue of hailstorm.*

According to a respondent (61-year-old, male),

4-5 *sālo se lage hai yahān* anti-hail nets. Anti-hail net

> *zyādā achchhe hai* anti-hail gun *se, magara tuphāna ājāye to* damage *bhī karate hai* net. 2011 *me* Bhuilaghat *aura* Deorighat *me* anti-hail gun *lagī aur* 2015-16 *me* Baghi *me. Jahān jahān* gun *lagī, vahān ke āsa pāsa ke* area *me olāvriṣṭī* double *ho gaī.* Radar *lagā hua hī nahīn hai* anti-hail gun *me. Tukke vālā kām hai. Bādala dikhā chalā lo.* Radar *batata olā vālā bādala hai ya bāriśa kā.* Khadapathar *vālā* radar *kharāba ho gayā hai. Nā bāriśa ātī nā olā rūkatā. Pichhale* 4 *sāla se garmī ke bhī bure hāl hai.*
>
> Anti-hail nets have been installed for the past 4-5 years. Anti-hail nets are better than anti-hail gun, but during a storm they cause damage too. Anti-hail guns were installed in Bhuilaghat and Deorighat in 2011 and in 2015-16 in Baghi. Wherever guns were installed, the areas nearby them suffered from twice the hailstorm. There is no radar in the anti-hail gun. Job is based on guess. Saw a cloud, fire it. Radar tells you if a cloud is of hail or rain. Radar of Khadapathar does not work. Neither it rains nor the hailstorm is stopped. Temperature has also been increased from past four years.

Though there was not severe issue of water scarcity, the people have started practicing rainwater harvesting. The respondents were of the opinion that this was an essential step that would help them in their future endeavours. The people showed an interest in new sources of income like tourism as they have observed an upward shift of the apple belt and felt that soon they would too, face the same issues as the lower belt villages.

According to a respondent (37-year-old, male),

> *Seba aba* risky *hai.* Second option tourism *dekha rahe hai hama.* 100 *me se* 5-10 per cent second option *dekha rahe āne vālī* generation *ke liye.* Cash crop *ke* survival *ke liye* reforestation *kara rahe. Pūrī panchāyat hara sāla peṛa lagā rahī kyonki hamare time me aba hame dikkata ā rahi. Aba āge vāle hamāre bachche to bacha jāye. Abhī pānī kī itanī dikkata bhī nahīn hai. Para agara* future

> *me* tourism *kā dekhanā hai to abhī se dhyāna rakhanā paṛegā.* Rainwater storage tank *bānā rahe. Jaba* source of income *nahīn rahegā* future *me to kuchha nahīn kara sakate. To abhī kara lo* invest. *Ādamī hara eka socha rahā ke* second source of income *honā chāhiye.*
>
> Apple is risky now. We are looking at tourism for second option. 5-10 per cent out of 100 are looking for second option for the coming generation. Reforestation is being done for the survival of the cash crop. The whole panchayat is planting trees every year as we are facing problem during our time. Now at least our children are spared in future. Right now, there is not so much problem of water. But if one has to look at tourism in future, care has to be taken from now itself. We are making rainwater storage tank. When there will be no source of income in future, we will not be able to do anything. So, invest now. Every man is thinking that there should be a second source of income.

The researcher observed that some respondents had table fans in their house, which implied that the temperature had increased considerably. The narratives indicate that the people of Jashla, even though were not facing any severe issue related to water, were practicing rainwater harvesting keeping in mind the future. They realised that sustainable use of natural resources like water was the need of the hour.

6.1.5 Changing dynamics of social relations as an adaptation to combat climate change

Social relations play an important role in livelihood management in the field area as *āpasadārī* (brotherhood, communal bond) was very strong in older generation. The respondents of younger generation believed that it was this *āpasadārī*, that was affecting their efforts to combat climate change. The strong communal bond between the household heads of the villages had both a positive and negative repercussion, while attempting to combat climate change. The respondents viewed that strong feeling of oneness and unity on one hand helped during mitigation steps like plantation drives for afforestation, the same feeling prevented them from stopping any activity that may adversely impact the climate. According to a respondent (35-year-old, male, from the lower-middle altitude village, Kiari),

> *Pahalī* generation *bigāṛanā nahīn chāhatī āpasa me, koi peṛa kāṭa rahā to kāṭane do, hame bhī zarūrata paṛegī.* But *ab* young generation *ke hātha me* command *ā rahī hai to koi peṛa kāṭe to* step *le letī, bulāo Paṭavārī ko auar roko kāma. Jaba hamare liye kuchha* scope *nahīn rahegā to kuchha to karana paṛegā. Hamein āge kā bhī to dekhanā paṛegā.*
>
> Older generation does not want to ruin the relations, someone is cutting tress then let them cut, we will also need. But now that young generation is getting the hold of the command in their hand, if someone cuts trees they take step, call the Patwari and stop the work. When there will be no scope left for us then we will have to do something. We will have to look at the future too.

Another respondent (42-year-old, male, from the upper-middle altitude village, Kufar Bag) recalled,

> *Kuchha samaya pahale kī bāt hai, sāmane vālo ne dekhā ki rāta ko* jungle *me koi peṛa kāṭa rahā ūpara vālī* side. *Tabhī* call *ghumāyā* 1090, Hoshiyar Singh Helpline *para. Agale dina ā gaye* forest *aphasara puchhane logon ko ke kauna thā peṛa kāṭane lagā.*
>
> Some time ago, the people in front of us saw that someone was cutting trees at night in upper side. They dialled then and there 1090, the Hoshiyar Singh Helpline. The next day forest officers arrived and questioned the people about who was cutting the tress.

Similarly, there have been incidents where water misuse by the people have led to fights. The respondents expressed that everybody should use the allotted water in a proper way so as to lessen the problem of water scarcity. These narratives are indicative of a shift in the dynamics of social relations, especially among the younger generation. The respondents justified this weakening of communal bond as a necessity to avoid the future and more adverse implications of climate change. This shift has come around in order to prevent the further degradation of the forests and thus the climate.

6.1.6 Steps to mitigate the impact of climate change

The people of the study area started afforestation as an effective step to mitigate the impact of climate change. The people have been participating in plantation drives along with various organizations such as Mahila Mandal, Mandir Committee and Ghyal Paryavaran Samiti. According to a respondent (61-year-old male, from the lower-middle altitude village, Kiari),

> Religion *kā* main role *hai. Koi Pradhān banake kuchha nahīn kara paegā kyonki tānga kāṭane ke liye saba hai. Para Devatā kā kehanā saba mānate hai. Hama se zyādā* new generation *chalā rahī Deva Nītī aur āne vālī* generation *aur* better *hogī. Jo loga ghara me jūṭhe* glass *nahīn dhote vo vahān* physical work *kara rahe. Hamane Deva Nītī ke ādhār pe kītane hi peṛa lagavāye hai* jungle *me. Aura loga āte bhī hai auar pūre mana se kāma bhī karate hai.*
>
> Religion has a main role. No one will be able to do anything after becoming Pradhan because everyone wants to pull the other down. But everybody listens to the deity. More than us, the new generation is keeping up the *Dev Niti* and the coming generation will be even better. The people who do not even wash their own used glasses at home, they are doing physical work. We have planted so many trees in forest on the basis of *Deva Nītī*. And people come as well and work whole-heartedly.

The respondents believed that religion played a crucial role in combating climate change as the people of Kotkhai tehsil were ardent worshippers of their local deities and it was only aspect of their society that had not changed over the period of time. All the people, older or younger generation believed in the *Deva*

Nītī, the command of their deities, *Grama Devatā, Iṣṭa Devatā* and *Kula Devatā.*

The present study confirmed that various adaptations were made in the form of adopting new methods like cultivation of High-Density Plant (HDP), installation of anti-hail net and gun, and harvesting of rainwater in order to combat the ongoing climate change. These findings concur with that of Datta (2013), according to which the effective way to combat climate change is to adopt conservation agriculture, use renewable energy, conservation of water, reforestation, etc. in order to minimize the impact of climate change and present horticultural practices need to be modified. Introducing new varieties of horticultural crops that are tolerant to high temperature, resistant to pests and diseases, provide high yield under stressful conditions, and do not exhaust the natural resources are the main strategies to combat the impacts of climate change.

Adaptation strategies are being used at household level as well as horticultural level, like conservation and conscious use of water for daily needs, and opting for better adapted varieties for horticulture purpose. These adaptations are in accordance to those observed by Vidaurre de la Riva *et al.* (2013), in their study on Charazani of Andean region of Bolivia where people have taken up several adaptive measures at household, community and external levels in order to cope with the events of climate change. The people used traditional knowledge and contemporary practices in agriculture and are also looking up for non-agriculture related activities to support the subsistence.

It has been identified in the field area that the respondents have expanded their source of income by taking up different jobs, like that of construction and builder job, and have started using vehicles for commercial purpose. This finding is in convergences with the study by Belachew and Zuberi (2015) of the farming community of Maruf Kebele, Central Oromia (Ethiopia) which states that the farmers diversify their livelihood options as an adaptation to irregularities in the local weather and adopted a number of steps to modify agricultural practices.

Thus, the present research demonstrates adaptation strategies such as use of new methods for horticultural practice, adaptation at household levels, expansion of livelihood options and taking up steps like afforestation for mitigating climate change. These inferences are in agreement with the work of Le and Vo (2021), they observed that the people living in the Mekong Delta had four types of livelihood adaptation to combat changing climate. These were changes in structure of agricultural systems, change of employment sites, relocation and out-migration.

As stated by Barnosky *et al.* (2012), it is important to understand the degree and impact of climate on livelihoods and indirectly, adaptive capacity at regional and local scales as almost 50 per cent of the earth's natural ecologies face similar challenges.

6.2 The government efforts being taken to combat and mitigate climate change

The government of Himachal Pradesh have been taking multiple steps to combat climate change. Projects and schemes are being implemented along with various policies in order to conserve the environment and mitigate the challenges of climate change, for instance,

- A State Centre on Climate Change has been established under the aegis of the State Council for Science Technology & Environment, Department of Environment, Science & Technology, Himachal Pradesh in fulfilment to the National Action Plan on Climate Change (NAPCC) with objectives to enhance ecological sustainability, to coordinate with different institutions to take up theme based specific research on climate change and its impacts in the state, to formulate policies and give inputs to the state government in the field of disaster management and climate change, etc. (http://www.hpccc.gov.in/Default.aspx)
- Japan International Cooperation Agency (JICA) assisted "Project for Improvement of Himachal Pradesh Forest Ecosystems Management and Livelihoods" is being implemented in six districts, i.e., Kinnaur, Shimla, Bilaspur, Mandi, Kullu, and Lahaul and Spiti (2018-2028). The aim of this project is biodiversity conservation and betterment of ecosystem services in forest area for sustainable development in Himachal Pradesh with main focus on

water, soil and biodiversity conservation and supply of raw material for the livelihood of local population. (https://jicahpforestryproject.com/about/rationale)

- The schemes like "Parvat Dhara" scheme have been formulated to conserve water resources in the state. The scheme places emphasis on rejuvenation of water sources that are on the threshold of extinction. Forest Department have started constructing storage structures in order to provide irrigation facilities in sloping fields under this scheme. (http://himachalpr.gov.in/PressReleaseByYear.aspx?Language=1&ID=21757&Type=2&Date=02/05/2021)
- Plantation drives are being carried out under the guidance of HP Forest Department every year where, as per norms for afforestation purpose, 1100 trees are planted per hectare of land. (Government of Himachal Pradesh, 2021)
- Government has launched 'Jal Se Krishi Ko Bal Yojana' scheme to combat plummeting natural flow of water in wake of climate change. Under this scheme check dams and ponds will be constructed in order to recharge ground water table and to ensure water availability throughout the year. (Government of Himachal Pradesh, 2021)
- Himachal Pradesh Mid-Himalayan Watershed Development Project was started in 2005 with the objective to reverse the process of degradation of the natural resource base and to improve the productive potential of natural resources and incomes of the rural households in the project areas. The stakeholders of this project are the Department of Forests, local gram panchayats (GPs), and the World Bānak. (https://www.greengrowthknowledge.org/project/himachal-pradesh-mid-himalayan-watershed-development-project)
- Rainwater Harvesting Structures (RWHS) like check dams, rooftop RWHS and water harvesting pits are to be created under National Water Mission's (NWM) campaign "Catch the Rain" with the tagline "Catch the rain, where it falls, when it falls". The campaign includes removal of encroachments and de-silting of tanks to increase their storing capacity, and removal of hindrances in the channels

which bring water to them from the catchment areas with the help of the people.

- To save crops from hailstorms, the HP State Government has started "Krishi Utpadan Sarankshan Yojna" (Anti Hail Net) scheme. Under this scheme, the State government will provide 80 per cent subsidy to farmers on the purchase of anti-hail nets to protect horticultural crops especially Apple from hailstorms. (Government of Himachal Pradesh, 2021)
- Funds have been allotted by the government of Himachal Pradesh for lift drinking water supply scheme. (http://himachalpr.gov.in/OneNews.aspx? Language =1&ID=18224)

6.2.1 Schemes and projects being implemented in the field area as per the respondents

Though there are numerous schemes developed by the government to combat and mitigate climate change, only five were being currently implemented in the study area. According to the respondents Watershed Development Project was started in the area in 2005. The project covered 9 panchayats in the dry belt from Purag (Bhawana) to Baghi.

Lift Water Supply Scheme (LWSS) by Irrigation and Public Health (IPH) department have been implemented in the field area. As per the Parala Badruni panchayat member (of Kufar Bag), funding of around rupees one crore five lakh was being provided for the scheme's proper implementation. The member of Kiari panchayat informed that Kiari and surrounding villages were availing the lift scheme. Water from Kangi Nala was being provided to Kiari and surrounding villages. Water from Giriganga was being supplied from Kotkhai (town) to Neraghati. Kufar Bag and Jashla village were also supplied water under LWSS from Kangi Nala.

Japan International Cooperation Agency (JICA) assisted "Project for Improvement of Himachal Pradesh Forest Ecosystems Management and Livelihoods" will be employed in the field area. As per a respondent (44-year-old, male, from the upper-middle altitude village, Kufar Bag),

> *Ek* JICA project *āyā hai.* Kufar Bag, Chol, Badruni *aura* Jakrari *me* street lights *lagegī* afforestation *hogī aura* 10 *saal ke liye pānī kī* supply *milegī.*
>
> One project has come, JICA. Streets lights will be installed, afforestation will be done and water supply will be arranged for 10 years for the villages Kufar Bag, Chol, Badruni and Jakrari.

Plantation drives are being led in the whole area under the guidance of Forest Department. People of Jashla and Kufar Bag had recently planted trees in the surrounding forest. The people have also availed the scheme of "Krishi Utpadan Sarankshan Yojana" in order to cut the expenses related to anti-hail nets.

Rainwater Harvesting Structures (RWHS) like water storage tanks (Plate 6.6) were being built at different places in order to harvest rainwater.

Source: Field data.

Plate 6.6: Rainwater harvesting tank.

Photo description: *The tank was built under the National Water Mission's (NWM) campaign "Catch the Rain".*

6.2.2 Issues related to the schemes faced by the respondents

The respondents observed that even though schemes have been launched for combating and mitigating the impacts of climate change, but effective implementation was yet to be done. A respondent (51-year-old, male, from the lower altitude village, Bhawana) stated that,

> Watershed Development Project start *to kīya thā para koi* proper use *nahīn huā. Is* project *me* check dams *bānaāī jānī thī aura* afforestation *honī thī. Nīche* catchment area *me pānī ātā aura phira vo* supply *hotā. Vo* land government *kā hai jo āja kī* date *me* encroached *hai.* Gumma to Baghi, 9 *panchāyata me ye* project *āyā thā, jo* dry belt *hai.* Soil conservation *bhī honī thī* by retaining soil by crate wall and check dams. *Para jo chhoṭe chhoṭe nāle hai vo to* encroached *hai to phira itna phāyadā huā nahīn jitnā honā chāhiye thā.*
>
> Watershed Development Project was initiated but it was not used properly. Check dams were to be built and afforestation was to be done under this project. Water would come in the catchment area below and that would have been supplied. That land belongs to the government which in today's date have been encroached. The project was implemented in 9 panchayats from Gumma to Baghi, which is a dry belt. Soil would have been conserved by retaining of soil by carte wall sand check dams. But as now small small streams have been encroached, the benefit was not as much as was expected.

In reference to the plantation drive, a respondent (63-year-old, male, from the lower-middle altitude village, Kiari) complained,

> *Peṛa ke liye gaye the hama* forest *vālo ke pās to bolate* registered *sansthā chāhiye peṛa lene ke liye. Achchhe kām ke liye chāhiye para nahīn denge.* Government *ke kām aise hī hai. Mān lo khuda se lagāne hai to nahīn denge. Bolate* panchayat, school *vālo ko yā kisī sansthā*

> *ko hi denge. Peṛa kāṭane ke liye koi* registration *nahīn, lagāne ke liye* registration *chāhiye.*
>
> When we went to the Forest Department people for trees, they told us that one needs to be a registered organisation for getting trees. We need them for good work but they would not give. Government work is like this only. Say we have to plant ourselves; they will not give. They said that they would give to panchayat, schools or any organisation only. There is no registration for cutting trees, registration is needed for planting.

Most of the respondents complained about the water supply under lift scheme. The water was supplied after 14-15 days in the area as per the respondents. A respondent (62-year-old, female from the upper-middle altitude village, Kufar Bag) expressed her view,

> 15-15 *din bād pāṇī ātā hai* Kangi Nala *kā.* Complaint *kī thī. Kuchha dina thīka āyā, uske bāda phira se vahī hāla.* Gutter *kā pānī hai to thīk hai, para agara bāriśa bhī nā ho to kahān jāyenge.*
>
> Water from Kangi Nala comes after 15-15 days. We filed a complaint. It came properly for few days, after that the same condition. It is okay if we have water from gutter, but if it does not even rain then where will we go.

The respondents informed that even though the government was providing budget for rainwater harvesting, and tanks were being made, it was useless as there was no supply of water and the tanks that were built were incomplete and open, thus exposing it to dirt and garbage. The people were building their personal tanks for storage of water.

Also, in spite of "Krishi Utpadan Sarankshan Yojana" being availed by many people of the field area, not everyone could reap its benefits as the subsidy was not being provided by the government on time. The people had to wait for it and also the whole procedure was time consuming and cumbersome.

According to some respondents, the forms perplexed them. A few respondents cannot afford the nets without subsidy and have to wait, which may be detrimental for the crop as it may be damaged by hailstorm in the meantime.

The respondents found that the forms to be filled to avail a scheme were complicated and financial aid took time to be granted. For instance, the people with low income were not able to afford the anti-hail nets without prior subsidy grant from the government. Therefore, education and economic conditions are the important factors that hinder the effective and appropriate response to the schemes. The successful implementation of these remedial measures by government can be achieved through generating awareness among the people.

7
Summary and Conclusion

7.1 Summary

Climate change has become a global threat due to its accelerated rate and deleterious effects on our planet and its inhabitants. Attempts are being made at global and local level to understand this phenomenon, the causative agents behind it, its impacts, adaptations required to mitigate it, and policies and strategies needed to combat it.

As per Intergovernmental Panel on Climate Change (IPCC, 2007a), "climate change refers to a change in the state of the climate that can be identified by changes in the mean and/or the variability of its properties, and that persists for an extended period, typically decades or longer. It refers to any change in climate over time, whether due to natural variability or as a result of human activity". These properties can be variables like precipitation, temperature, humidity, wind, and severe weather events.

Livelihood is defined as the capabilities, assets (both material and social resources) and activities that are required for a means of living and it is sustainable so that it can cope with, and recuperate from, stress and shocks, preserve or improve its capabilities and provide livelihoods opportunities for both present and the next generation without undermining the natural resource base (Chambers and Conway, 1991).

The present descriptive and exploratory research has been carried out to understand climate change as perceived by the people of Kotkhai tehsil of District Shimla (Himachal Pradesh), and its impact on their livelihood. The traditional strengths of the field of anthropology are participant observation and ability to give insight into the perception and knowledge of the people. The emic viewpoint in the form of narratives have been used to comprehend the reality of climate change at local level. An effort has been made to document climate change and its impact by taking into consideration not only the perception of people and local knowledge, but also using the scientific data related to climate conditions available, and comparing it with the local knowledge.

The unit of study for the present research are the households of the four villages, viz., Bhawana, Kiari, Kufar Bag and Jashla of Kotkhai tehsil. These villages are selected on the basis of their relative altitude. The study attempts to assess climate change by analysing the available climatic data for the region, i.e., temperature, rainfall and snowfall. The impact of these changes on the various aspects of livelihood of the people have been explored using a holistic anthropological approach.

The sample comprises of 97 households which have been selected after applying inclusion and exclusion criteria. A mixed method approach, i.e., both qualitative and quantitative methods of data collection have been used. As the present study is anthropological in nature, an emic view of the people has been presented as an analysis. The primary data is qualitative in nature and has been analyzed by using case studies and findings of Focus Group Discussions (FGDs) as narratives which has been transcribed using the system of transliteration (Appendix). The secondary data, being quantitative in nature, are analyzed using statistical techniques with the help of Microsoft Excel (MS Excel) and Statistical Package for the Social Sciences, version 16 (SPSS 16 Incc.) Triangulation of the data is done in order to enhance the validity of the data.

It has been found during the analysis of the data that both respondents' perception and the climate data reveals an increasing trend in temperature and a decline in rainfall. The rainfall pattern has been inconsistent as per the respondents as well as the data provided by IMD. The linear regression trendline shows an increasing trend for temperature and post-monsoon rainfall. The annual mean rainfall documents a declining trend with linear regression trendline presenting a downward trend for pre-monsoon, monsoon and winter rainfall. These results support the perception of people regarding shift in the months of monsoon rainfall from July-August towards August-September.

The trend in snowfall based on climate data provided by IMD does not match the perception of respondents. According to the respondents, snowfall has displayed a drastic change, both in amount and in its seasonality (timings). The increasing trend presented by the statistical analysis is based on the data of eleven years alone and it would not be out of place to mention that the amount of snowfall in the year 2020 has been maximum in the decade, which may have affected the trend. Also, the statistical data is cumulative in nature and gives a general trend based on the recorded data. However, the extent of climatic parameters' effect has shown variability according to the altitude as per the data collected by the researcher from the field area.

The impact of climate change on day-to-day activity is perceived differently by females and males of the study area. The men did not observe any major changes in daily activities, except for increase in work required in the orchards to combat the challenges of changing climate. They are more concerned about the long-term impact of changing climate on their source of livelihood, horticulture. The women on the other hand feel that climate change impact like scarcity of water has affected their daily activities because household chores are dependent on water. Other than the household chores, women have trouble with keeping up the hygiene regime during periods, also, they are often the victim of the outburst of men of the family who are stressed due to changing climate's impact on their source of income.

The present study documents the changes in crop varieties, crop production, their cropping time and their shifting calendar due to climate change. Kotkhai tehsil have different varieties of crops being grown at different altitudes. In the vilages Bhawana, Kiari and Jashla, people cultivate both pome and stone fruits, i.e., apple, pears, cherry, plum, peach and apricot. While in Kufar Bag people only cultivate pome fruit, apple. Climate change has affected the type of fruit crops that are being cultivated at the field area. New varieties of crop, like Gala and spur variety of apple, are being produced along with the traditional varieties.

The crop calendar is dependent on the altitude of the village and the climatic conditions such as the amount and time of snowfall and rainfall, and whether the temperature requirements are being met. Out of the total respondents of the selected villages, 49 per cent found a decrease in chilling period due to declining snowfall, 37 per cent stated that the chilling period is non-uniform due to variation in timing and amount of snowfall, and 14 per cent viewed the chilling period to be the same over the years. About 48 per cent respondents responded that the flowering started earlier, while 52 per cent observed it to be the same. All the respondents observed a delay in harvesting time regardless of the flowering time.

The respondents at Bhawana (the lower altitude village) and Kiari (the lower-middle altitude village noticed a trend of early flowering in plants and a delay in harvesting time due to negligible amount of snowfall in recent years. While in upper-middle altitude village, Kufar Bag, flowering remained unaffected, and in higher altitude village of Jashla, the flowering and setting depended on the timing of the snowfall and temperature conditions. For instance, late snowfall resulted in late flowering and delay in overall process of development of the fruit which further delayed the harvesting time.

Changes in pre-flowering stages, flowering, fruit set and fruit development stage, fruit growth and development stage, and harvesting time have been observed in the field area mainly due to change in climatic parameters over the time and also due to irregular pattern of climate. Overall, the harvesting stage has been delayed in the entire study area.

Both people's perception and data provided by the Department of Horticulture (Shimla) have been taken into account while analyzing the crop production. According to the respondents, the crop production has not been uniform. The productivity has been perceived to decline over the time by all the respondents. The general increase in production is due to increase in the area and cultivation of new varieties of crop which provides good results within small period of time. Also at higher altitude, as the weather is getting warmer, the areas that used to stay under snow cover and were not suitable for any crop production, are now turning into suitable options for production of apple.

Snowfall plays the most important role in apple and pear production.The warmer temperature favours stone fruits like peach and plum as they require fewer chilling hours. The production has been different for different altitude villages, but the productivity has been declining at lower altitude areas and is non-uniform at higher regions. The respondents have observed a shift in apple belt from lower altitude to higher altitude areas due to unfavourable climate conditions in the lower regions. The climatic conditions for the production of stone fruits are favourable at lower altitude area.

According to the data provided by the Department of Horticulture (Shimla) for Kotkhai tehsil (2007-2017), apple is the most widely grown fruit crop in the area and comprises the largest portion of the crop being produced in the Kotkhai belt, followed by pear, cherry, plum, apricot and peach. Maximum productivity of apple has been documented in the year 2011 and minimum in the year 2012. During the year of maximum productivity (2011), highest snowfall and summer rainfall were recorded. While in the year 2012, the year of minimum productivity, least amount of snowfall and summer rainfall was observed. This result synchronizes with the perception of the people that the snowfall and summer rains are important for apple productivity.

Maximum productivity for pear was observed in the year 2010 and minimum in the year 2009. The productivity did not exhibit any relation to the climatic parameters. The maximum amount of snowfall for the year of maximum productivity (2011) for plum does not coincide with people's perception that warmer conditions and fewer chilling hours are required for plum productivity. Data for peach indicate towards maximum productivity in year 2009 and minimum in year 2012. The high maximum and minimum temperature, and less winter rainfall during the initial growth phase, has been more for the year of maximum productivity, thus, complementing the perception of the respondents that warmer winters were favourable for stone fruits than pome fruits.

Apricot documented maximum productivity in year 2011 and minimum in year 2010. The maximum and minimum temperature during the initial growth phase has been more for the year of minimum productivity (2010). Highest snowfall was recorded in the year of maximum productivity (2011), thus, opposing the perception of the respondents that stone fruits productivity is not affected by lesser chilling hours and warmer conditions.

Maximum productivity for cherry was observed in year 2008 and minimum in year 2015. There has been no rainfall during the different growth phases for the year of maximum productivity (2008). The climatic data neither demonstrate any relation with the minimum productivity nor with the people's perception.

The secondary data depicts varying relationship between climatic conditions and productivity. The findings from the analysis of data provided by IMD and Department of Horticulture does not completely converge with the perception of the people. It is pertinent to mention that the crop production is not solely dependent on climatic conditions, but also on the various steps taken to enhance the production, gradients like altitude and whether the area is sun-facing (south facing slope) or lies on the shady side (north facing slope) of the mountain.

The impact of climate change on the different assets of livelihood has been observed by 98 per cent of respondents of all the selected villages. Of these 98 per cent respondents, different extent of impact has been observed on the livelihood assets. Maximum impact has been observed on the natural assets, followed by human, social, physical and financial assets.

The respondents perceived natural resources to be the worst affected assets of livelihood which has adversely impacted their day to day lives and source of income, which is primarily based on horticulture. Due to increasing temperature and decline in snowfall and short-lived monsoon, the land has been losing its moisture content and is becoming dry. The warmer winters does not allow the snow to stay for longer duration, hence leaving less time for seepage of water. The slope of the hill on which the villages were situated also affected the moisture of the land. The respondents from the villages that are in *parālā* (on south facing slope, sunny), Kiari and Jashla, observed that snow stays for lesser time now as the temperature is increasing. The respondents from Kufar Bag (*śīlā*, shady as lying on north facing slope) informed that the snow lasts for longer time in the village which leads to moisture retention in the soil. While in Bhawana, respondents stated that even though the village lies on a south-east facing slope, i.e., on both sunny and shady part of the hill, due to negligible snowfall since past few years, the land has dried up. Also, all the villages faced the issue of untimely lashes of rainfall which hardly percolated the surface and washed away nutrients along with it as surface run off. Thus, impacting the nutritional content of the soil.

Water as a livelihood asset has been one of the most adversely impacted natural asset in the area. The respondents believed that increasing temperature, minimal snowfall, and inconsistent and untimely rainfall has dried up almost all the natural water sources in the area and there was a significant decline in the groundwater level. The most striking impact on water in the lower and lower-middle altitude villages was the disappearance of *chaśamā* or *moka* (aquifers) which were the traditional natural source of water in the area. These areas were now dependent on the little water left in the *khaḍa* and *bāoṛī* respectively and lift scheme water that was supplied by the government through channelling water from water sources in the higher altitudes.

The respondents from Bhawana (the lower altitude village) observed that the natural water sources have dried out. Local sources of water have completely depleted since the year 2000. The water that was available in their orchards, in *bāoṛī* is finished now. Similarly in the lower-middle altitude village, Kiari, the respondents observed that over the years, the natural water sources have dried up due to decline in snowfall and rainfall. While in comparison to Bhawana and Kiari, there are natural water sources available in upper-middle altitude village, Kufar Bag, though with time these sources (*chaśamā* and *kuphara*) have considerably depleted. In Kufar Bag, the extent of issue of water scarcity increases from bottom to the top of the hill on which the village is located on. The respondents living at the top of the village suffered considerably from water scarcity in comparison to the respondents living at the lower part of the village.

In Jashla, the high-altitude village, the respondents have not faced much issues related to the water. However, a decline has been observed in the quantity of water in the available sources. Recently, lift water scheme has also started in the area and rainwater harvesting is being done by people living in lower parts of the village.

Climate change have impacted the biodiversity in the area. Species of plants (*kāfala*, blue berry, *jāmuna* and *banaphaśā*) and animals (*barāgha* and *rīkha*) that were found in the area in the past have become rare and new invasive species of grass have emerged in the study area. The forest has become thinner due to deforestation and forest fires. The new trees being planted cannot cope up with the existing climate conditions resulting in failure of plantation drives and weak, thin trees.

Forest fires have raised concern in the area. Respondents are of the view that increase in temperature and lack of moisture in soil, made it easier to catch fire in forests. Grassland fires have equally been a cause of concern in the lower belts due to scorching heat and lack of moisture.

The respondents noted that changing climate has affected the household, as members, especially the younger generation, were moving out as the future of horticulture was climate dependent and hence, uncertain. A trend of out-migration was observed among the younger generation for better education and job opportunities.

The active labour has been affected both in context of time and workload. Due to extreme weather events like excessive heat, erratic lashes of rainfall leading to cloud bursts and hailstorm, it has become difficult to work in the field. The people have become 100 per cent dependent on the labourers. The labourers, which are the Nepali people, often face problems due to such climatic conditions as they do not have proper houses to live in, thus making them hesitant to work in areas that are prone to such events. Also, the labour wages have increased over the time and once they have enough money, they go back to Nepal for better education for their children and starting their own work.

The health of the people has been affected by the changing climate as per the people. The climate change has impacted the physical, mental and social health of the people of the study area as perceived by the respondents. Some respondents noticed an immediate impact of climate change on health, while other noted that the changing climate would have implications on health in near future.

The respondents are of the belief that due to changes in climatic conditions there has been a surge in different types of diseases related to plants and animals. There has been an increase in the use of pesticides and chemical fertilizers for promoting the growth of various crops as well as protecting them from various emerging diseases, and unfavourable climate conditions which have led to issues like lack of moisture and nutrients in soil, and excessive moisture or drought like conditions. These chemicals have entered into the nutrient cycle and water cycle which have impacted the health of the people. The respondents believed that the older generation were healthier and stronger in spite of no medical facilities than today's generation as they lived in a clean environment and ate healthy, organic, chemical free food that was grown at home.

According to the respondents, the increasing temperature had made it difficult for people to work in fields during the day time. If proper precautions are not taken, one might suffer from heat stroke. The scarcity of water due to shortage of annual snowfall and rainfall have affected the health directly by contaminating the water used for drinking, daily hygiene and sanitation, and indirectly through their livelihood based on horticulture. The reason for decline in health of children and elderly may be attributed to the quality of air which is deteriorating due to incidence of forest fires, which are becoming frequent in the forests nearby due increase in temperature and drying patches of land and grass owing to the lack of moisture and increased heat. Also, burning of garbage also leads to respiratory problems like lower vital capacity and allergies as it comprises of toxic waste material, such as tetra packs and containers of chemicals being used in the fields for various diseases related to our crops, along with household waste.

Shortage of water have affected apple production, which is the main source of livelihood for the people in the area. The changing and erratic weather conditions have led to stress and anxiety among the people.

The migration of younger generation from villages to the cities has led to a sense of loss in the older people who are left behind in the villages as there has been an influx of new cultural traits and ways of life as the younger people are coming in contact with people of various culture and are imbibing their traits which has led to acculturation, mostly in case of interstate marriages. The elders have developed a sense of isolation and grief as there was so much changing around them– the temperature, the precipitation, the way different seasons felt, and the culture and their communal bond.

The respondents observed that as people are yet to adapt to shortage of water, fights generally broke out amongst the villagers regarding the use of water, thus, affecting harmony of the village. The respondents have faced incidents like fights over water, which is hampering the peace and smooth functioning of the village community by causing tensions between the people and affecting the communal bond, and increasing the cases of interpersonal grudges.

The respondents believe that new sets of skill and knowledge are to be learned while preserving the traditional knowledge, for instance, traditional knowledge regarding *kuphara* (natural check dam), *Kaniṣka* (*Alnus nepalensis*), *bāna* (*Quercus leucotrichophora*) and *moharū* (*Quercus dilatate*) trees. Also, skill like rainwater harvesting and knowledge about various diseases emerging due to climate change is the need of the hour. The respondents in the field area have started using anti-hail net and anti-hail gun to combat the impact of hailstorm. The respondents had knowledge about the new disease emerging in the area and have started using fertilizers and pesticides as per the requirement. The area has seen a widespread use of High-Density Plants (HDP). These comprises of spur and gala varieties of apple. According to the respondent, they are better suited for warmer climate as they needed fewer chilling hours.

The respondents claim to have the strong *āpasadārī* (brotherhood, communal bond), community or group involvement in activities and strong socio-religious influence. Impact of climate change on these aspects of the lives of the people has also been observed by the respondents. Some respondents observed gradual changes in customs associated to death and marriage over the time. Earlier during incidences like a death in the village, all the women of the village would get together and wash all the clothes, used and unused, in the house of the deceased irrespectively and clean the whole house. Now only women of same clan help in these chores and wash only those clothes that are used in order to save water.

Also, earlier ceremonies like marriage would take place at home and lasted for five to seven days where all the relatives and villagers would gather and share their joys and experiences. These days weddings usually take place in a hotel, for two days at the maximum, in order to avoid the trouble of cleaning afterwards as well as due to the busy schedule as everyone was engrossed in their work, be it orchards or any job. Not all people could attend these weddings as someone had to stay back at home for taking care of their land and domestic animals. Old people usually miss out these ceremonies leading to a sense of isolation among them.

During the study it was observed that the majority of population in villages comprised of old people, their children would only visit during holidays or during the harvesting of the crop, i.e., the apple season. The respondents believed that although the younger generation were still attached to their roots, their villages, the uncertainty of livelihood flourishing in near future due to climate change have forced them to move out, thus creating a social void. Respondents from older generation feel that there is a chance that the coming generation may forget their roots and communal bonds altogether. The most prominent social asset in the study area is that of socio-religious influence of the local deities. All the people have strong faith in their deities.

Climate change have impacted the practice of cattle rearing, which was a common practice in the older times. The changing climate which has led to emergence of new diseases, have resulted in an increase in use of chemicals in the form of pesticides that has contaminated soil and water. According to the respondents, this has contaminated the grass used for fodder and water used for drinking, which has impacted the cow's yield and quality. Also, as the villages comprises mainly of people of older generation, it has become difficult for them to manage cattle.

In recent times most of the people have stopped raising cows due to lack of water and are now dependent on milk in tetra packs, which in comparison to cows' milk have lesser nutritional aspect. This has led to increase in expenditure in the form of milk and *ghī* (clarified butter). Earlier the cow manure, used as fertiliser, was available at home. But now people have to buy manure, thus increasing the expenditure on horticulture.

Climate change has impacted not only cattle rearing but also bee-keeping business. The erratic weather conditions during the pollination season like sunny days and rain in the evening affected the productivity of the bees thus impacting the bee industry. Equipment like grass cutting machines is being used more frequently due to invasive new species of grass that grow at a very fast rate. Excessive use of these machines often leads to the damage to the blades and motors. The inconsistent extreme weather event like cloud burst or heavy rainfall have caused damage to various physical assets.

With the advent of cash crop economy in the form of horticulture, a boost in the financial status of all the land holders has been observed. With the climate change, the productivity may have gone down, but the production has still increased with the help of combating tools and techniques and an increase in the area of production. The people of lower and lower-middle altitude villages have also resorted to different source of income such as growing crops that are suitable for warmer conditions, owning commercial vehicles and taking up contract related to road development and construction work. Other than that, most of the households had a family member who had a job or was into business other than horticulture.

The main source of livelihood for the people of field area is horticultural produce, of which apple is the major crop produced, while fruits like pears, cherry, plum, apricot and peach are grown at a very small scale. Horticulture, being dependent on climatic parameters like temperature and precipitation, have faced a brunt of climate change. Natural assets of livelihood were found to be affected by changing climatic conditions which in turn affected horticulture practices which then affected the human, social, financial and physical assets. All the respondents felt an impact of climate change on fruit production.

Fruit industry depends on regular and timely cycle of seasons every year. Changes in climatic parameters has affected the fruit productivity and yeild. The respondents observed that storage and shelf life of the produce has declined as temperature has shown rising trend. As per the respondents, for proper colour development, low temperature is required just before harvesting. Due to increase in temperature, apple fails to change into desired colour of produce. Also, high temperature during harvesting season at times lead to sunburn in fruits and delay in leaf fall.

Cool temperature is vital in winter as chilling requirement of the fruit must be met. This chilling period during dormant phase has a crucial impact on bud break, flowering, development and therefore on the quality of fruit. Insufficient winter chilling period has severely affected crop yields in the field area as lack of chilling hours has led to irregular flowering leading to non-uniform crop development which affects the crop size and maturation in the same tree, thus reducing the yield.

Apple uses 100-200 litres of water per tree. A decline in rainfall has led to shortage of water to an extent that the daily requirement of an apple tree has become difficult to meet. The people have shifted to other means of livelihood such as jobs in other sectors. However, the youth who are still interested in their ancestral work are trying to come up with the solution, such as High-Density Crops in lower belts. A decrease in rainfall or delay in rainfall will affect the fruit quality, like lower juice content, smaller size, lower fruit weight and absence of colour. Slight changes in precipitation, its timing and duration, and intensity affects the fruit crop. Change in rainfall pattern like timing, intensity, onset and withdrawal of monsoon, cloud bursts, and dry spells affect the growth and development of the fruit and also give rise to diseases related to the fruit.

There is a decline in the apple production in lower belts as there has been little to no snowfall in that area. At the budding stage, apple production requires 1200 to 1500 hours of continuous temperature below 7°C between the months of January and February, but it is not met. Since past few years, hill chill varieties like apple have been switched with low chilling requiring varieties like High Density Plants (HDP), peaches, etc. As a decline in precipitation and increasing temperature, the apple in lower belt is shifting towards higher belts and are being replaced by low chilling requiring stone fruits.

Parts of higher altitudes where apple production was not feasible before have now become the perfect grounds for apple production. During the earlier times, the region used to stay under snow cover for a long period of time, exceeding the chilling period needed for plant growth, and thus making it impossible for the apples to grow there. Now, due to lesser snowfall and increase in the temperature comparatively to the earlier days, the optimal conditions for apple production are possible. Temperature, rainfall and snowfall, all need to be ideal for a good crop.

Climate change has resulted in unpredictable weather conditions. Sudden rainfall during the setting period or no snowfall during the chilling period has been a problem. A shift in seasons has been observed which makes it difficult for people and crop to adapt. According to people, when there is proper rainfall and chilling period requirement is achieved due to snowfall and low temperature during winter, the apple production is satisfactory. But in erratic weather conditions, there is a decline in apple production. So, depending on the amount of snowfall, rainfall and temperature, apple production may be good or poor, which has a direct impact on the livelihood.

The changing climate has led to change in epidemiology of pests and disease-causing agents. Excess of untimely rainfall and increasing temperature have led to different diseases or pest/insect problems. This has led to increase in the use of insecticides and pesticides. Due to climate change (less rainfall, higher temperature, erratic weather), people are using chemicals to sustain apple production and therefore increased expenditure. Also, in this way climate change is not only impacting livelihood, but also may have negative effect on human health as well as soil productivity and organisms living in it.

The people of the study area have acquired adaptation strategies to overcome the climate change. As the impact of changing climate is of different extent at different altitudes, adaptation strategies related to the major issues faced by the respondents of different villages were acquired in accordance to the impact observed.

In Bhawana (the lower altitude village), increasing temperature, minimal to no snowfall, and declining and erratic rainfall have the major bearing on their water sources and crop production. In order to combat the adverse effects of the changing climatic parameters, the respondents have incorporated various changes and tactics and are trying to adapt to their new setting. The respondents practice rainwater harvesting to collect water for daily use. Tanks are being built in order to collect rainwater. Water supplied by government sources is stored and used as per need. The people of the village are aware of water scarcity and make a conscious effort at individual level to not waste water. With the development of roads and easy access to market nearby, in Gumma, it has become easier to arrange milk and related products.

The respondents have shifted to stone fruits such as plum, apricot and peaches, and High-Density Plant (HDP) like spur and gala variety of apple from the traditional apple variety. The respondents who had flat land available have started growing HDP. It does not need long chilling hours as compared to the earlier variety of apples being grown in the area. Also, the roots of the HDP are shallow, so they absorb water from the surface and do not require deeper percolation of water. Thus, they are suitable for the declining rainfall and increased temperature in the area. The respondents are using mulching sheets and mats in order for better absorption of water. The people of the village felt that with changes in climate their main source of income, that is, horticulture was being impacted. New array of diseases and pests are emerging, therefore changes in the horticulture practices is required. The people are becoming more scientific in their approach. But more efforts are to be put in by the people to combat climate change. People who cannot invest much in HDP, feel that apple will not last long so they are shifting towards stone fruits for commercial purpose.

During the research it was observed that all the households have secondary source of income. The people are either working in cities or have taken up construction or builder job. People have bought land at places that are still not impacted as much as Bhawana due to climate change.

In Kiari (the lower-middle altitude village), the respondents felt that water scarcity is the major issue in the area. Therefore, rainwater harvesting is being done at household level. The people collect rainwater through gutters that have been installed on the roof. The water is used for orchards purpose as well as for household use. Conservation of water at individual level is being done.

Change in mode of transportation of apples has been observed. While earlier trucks were hired to transport apples to the market, due to decline in apple production people use vehicles like utilities. This has helped them in saving money of transport and also given employment opportunity at local level. Most of the people now own commercial vehicles as source of income in the village. Other than commercial vehicles, majority of the younger generation were found to be working in cities for more reliable source of livelihood.

The respondents felt a need to switch to HDP varieties of apple and stone fruits as they survived in warmer temperature and needed relatively less water than the earlier varieties of apple. The respondents are also trying their hand in organic farming in order to conserve the soil for the next generation.

Water scarcity and hailstorms are the main problem in Kufar Bag (the upper-middle altitude village). The people are practicing rainwater harvesting and rooftop gutter system are being used to harvest rainwater. The respondents believe that the traditional apple trees can only survive in the micro-climate of the village. High Density Plants (HDPs) need certain initial requirements, like gentle slope, easy access to road and drip irrigation facilities, which cannot be met with in Kufar Bag. Traditional trees are still working so there is no need either. However, plantation of forest trees is required to combat climate change.

Anti-hail guns have been installed in the surrounding areas. The respondents observed an increase in frequency of hailstorm after the installing of these anti-hail guns, first in year 2011 in Bhuilaghat and second in year 2015-16 in Baghi village. The respondents believe that due to lack of knowledge about the clouds that caused hailstorm, the people dispersed every cloud in fear of hailstorm. The respondents did not face much problem due to hailstorm before, however, after the instalment of the anti-hail guns in nearby areas, frequency of hailstorm increased. The hailstorm of year 2020 caused so much damage to the crop that some of the people installed anti-hail nets the next year, 2021.

The respondents were using new and effective chemicals to combat with the emerging diseases. Due to invasion of new species of plants in the orchards, it was becoming difficult for the people to manually clear the orchards frequently. The people have started using grass cutting machines for this purpose as it is less laborious. In Jashla, an anti-hail gun was installed in Deorighat in year 2011 in order to combat frequent hailstorm in the area. They have now installed anti-hail nets too as the anti-hail gun did not have the expected effect.

Though there was not severe issue of water scarcity, the people have started practicing rainwater harvesting. The respondents felt that this was an essential step that would help them in their future endeavours. The people showed an interest in new sources of income like tourism as they have observed an upward shift of the apple belt and felt that soon they would too, face the same issues as the lower belt villages.

The government of Himachal Pradesh have been taking multiple steps to combat climate change. According to the respondents Watershed Development Project was started in the area in year 2005. The project covered 9 panchayats in the dry belt from Purag (Bhawana) to Baghi. Lift Water Supply Scheme (LWSS) by Irrigation and Public Health (IPH) department have been implemented in the field area. Japan International Cooperation Agency (JICA) assisted "Project for Improvement of Himachal Pradesh Forest Ecosystems Management and Livelihoods" will be employed in the field area. Plantation drives are being led in the whole area under the guidance of Forest Department. People of Jashla and Kufar Bag had recently planted trees in the surrounding forest. The people have also availed the scheme of "Krishi Utpadan Sanrakshan Yojana" in order to cut the expenses related to anti-hail nets. Rainwater Harvesting

Structures (RWHS) like water storage tanks were being built at different places in order to harvest rainwater.

The respondents stated that even though various schemes have been made for combating and mitigating the impacts of climate change, proper implementation was yet to be done. Education and economic conditions are the important factors that affected proper use of the schemes. Policies made by government are accessible only if one is aware. The forms to be filled are complicated and financial aid takes time to be granted thus impacting the effective use of the policies and strategies formulated by the government.

7.2 Conclusion

The findings of the present study indicated towards an increase in the temperature, and decline in the snowfall and rainfall which has affected the livelihood. The impact of climate change on the livelihood has been perceived differently by the people living in the villages at the different altitude. The lower-belt areas have been found to be more affected by the changing climatic conditions, while the higher belt areas are also becoming vulnerable to climate change over the time. It has resulted in a decline in apple production in lower and lower-middle altitude villages. It has led to the problem of water scarcity in all the villages. In case of high-altitude village, the higher reaches are now available for apple production, thus increasing the source of income. Therefore, the findings of the present study indicate a shift in the ecosystem boundary.

The changes in climatic parameters have taken a toll over the natural resources in the whole area, of which the decline in water have had the major impact as it affected every aspect from day-to-day activities to requirement for horticultural practices. A difference in impact on the means of livelihood was seen when gender was put under the microscope. The women were facing more trouble in day-to-day activities as shortage of water meant less water for household chores. Men were not concerned with the issues of this level and were more troubled by the impact on the economy.

Various livelihood assets have been impacted by the changing climate. The soil content has deteriorated, water has become scarce, various species of flora and fauna have declined or vanished, and new invasive species have taken over. As the people of the field area were mainly horticulturists, the changes in climate conditions had led to anxiety and stress amongst the people due to the uncertainties in weather patterns and rise of new diseases which has resulted in use of chemical fertilizers and pesticides. There has been a trend of outward migration as the future of present source of income, horticulture, is bleak due to its dependence on nature and climate, leaving behind the older generation with a sense of loss and grief.

As far as the people who are still in the villages are concerned, there has been a change in social relations along with the shift of power dynamics from older to younger generation. While conducting the study it was observed that the *āpasadārī*, the brotherhood and the communal bond was kept intact by the *Deva Nītī*, commands of their deity.

The climate change in the form of increasing temperature and declining precipitation have jeopardized the health of people by compromising the quality of food and water; by leading to stress and anxiety amongst the people due to the uncertainty of the climate conditions; by developing sense of grief, and affecting the social health by changing the dynamics of the various social activities and interactions, and bringing in changes in the communal bond and affinity, thus disturbing the balance of social cohesion.

Even though advent of apple-based cash crop economy had boosted the financial condition of the people of the field area over the time, increase in expenditure on various physical assets, and decline in productivity of apple and shift from apple to stone fruits, have been observed due to climate change.

It can be concluded from the study that real cases are needed for the virtual reality to assess the ground reality of today's on-going hue and cry regarding impact of climate change. All the discussion regarding emission reduction amid government and private sectors has lost the focus from livelihood and actual people as it is more oriented towards industries, government and nongovernment sectors than towards people. This study attempts to understand how people perceive the change in climate, the reasons behind it and the impact on their livelihood.

A cause-and-effect relationship exists between climate change and livelihood. The people in order to increase the economy exploited the nature in the name of development. Over the time, due to changing climate, the precipitation decreased and temperature increased which led to a decline in apple production and an increase in expenses to combat the impact of the changing climatic conditions.

Climate change is a global phenomenon which needs action at global level, however, with small steps taken to combat it at individual and community level, contribution to deaccelerate its alarming rate can be made. The present accelerated climate change does not act in isolation, but in nexus to the livelihood related anthropogenic activities. The impact of climate change can be reduced only when actions are taken to combat climate change. What one needs to realise is that a sustainable approach is required towards use of resources, so that even when the resources are in bounty or shortage, one is able to live through it without depriving the coming generation of it.

Climate change has led to transitions in the way of life which needs action to be taken in order to adapt to these changes in climate. The people of the study area have taken up steps like afforestation and rainwater harvesting at individual and community level to combat the impact of climate change. An attempt to use organic fertilisers and pesticides has been made to counter the negative implications of using harmful chemicals.

The government have formulated various policies and strategies to combat climate change as is has become mandatory to deal with it. However, there is a need to assess whether these policies and strategies are being properly implemented and executed. Education and economic condition seem to be pre-requisite in case of the present research where people who were not educated were not either aware of the policies or were unable to access it due to the cumbersome form filling and formalities required.

Appendix

System of transliteration

Letter in Hindi	Transliteration in English
ँ /ं	n
अ	a
आ/ ा	*ā*
इ/ ि	i
ई/ी	*ī*
उ/ु	u
ऊ/ू	*ū*
ऋ/ृ	*r̲*
ए/े	e
ऐ/ै	ai
ओ/ो	o
औ/ौ	au
क	ka
ख	kha
ग	ga
घ	gha
ङ	*ṅa*
च	cha

Letter in Hindi	Transliteration in English
छ	chha
ज	ja
झ	jha
ञ	*ṣa*
ट	*ṭa*
ठ	*ṭha*
ड	*ḍa*
ढ	*ḍha*
ण	*ṇa*
त	ta
थ	tha
द	da
ध	dha
न	na
प	pa
फ	pha
ब	ba
भ	bha
म	ma
य	ya
र	ra
ल	la
व	va
श	*śa*
ष	*ṣa*
स	sa
ह	ha
ज़	za
ड़	*ṛa*
ढ़	*ṛha*

Example of transliteration

हमारे ज़माने में तो बारिश लगातार होती थी
Hamāre zamāne men to bāriśa lagātāra hotī thī

Source

गीतोपनिषद् Bhagavad-GĪTĀ AS IT IS
Second Edition
By
A.C. Bhaktivedanta Swami Prabhupāda
The Bhaktivedanta Book Trust, New Delhi

References

Adger, W.N., N. Brooks, G. Bentham, M. Agnew, and S. Ereksen. 2004. *New indicators of vulnerability and adaptive capacity* (Vol 7). Norwich: Tyndall Centre for Climate Change Research.

Ahmed, N., S. Lal, B. Das, and J.J. Mir. 2011. Impact of climate change on temperate fruit crops. In *Impact of climate change on fruit crops*. W.S. Dhillon, and P.S. Aulakh, eds. pp. 385-402. Delhi: Narendra Publishing House.

Ahuja, R. 2007. *Research methods*. Jaipur: Rawat Publications.

Alderman, H. 2010. Safety nets can help address the risks to nutrition from increasing climate variability. *The Journal of Nutrition* 140(1): 148S-152S.

Asch, T., and P. Asch. 2003. Film in ethnographic research. In *Principles of visual anthropology* (3rd ed.). P. Hockings, ed. Pp. 335–360. Berlin: Mouton de Gruyter.

Asian Development Bank. 2012. *Addressing climate change and migration in Asia and the Pacific.* Philippines: Asian Development Bank.

Baer, H.A., and M. Singer. 2009. *Global warming and the political ecology of health: emerging crises and systemic solutions.* California: Left Coast Press.

Barnes, J. 2014. *Cultivating the Nile: the everyday politics of water in Egypt*. Durham: Duke University Press.

Barnes, J., M. Dove, M. Lahsen, A. Mathews, P. McElwee, R. McIntosh, F. Moore, J. O'Reilly, B. Orlove, R. Puri, H. Weiss, and K. Yager. 2013. Contribution of anthropology to the study of climate change. *Nature Climate Change,* 3: 541-544.

Barnosky, A.D., E.A. Hadly, J. Bascompte, E.L. Berlow, J.H. Brown, M. Fortelius, W.M. Getz, J. Harte, A. Hastings, P.A. Marquet, N.D. Martinez, A. Mooers, P. Roopnarine, G. Vermeij, J.W. Williams, R. Gillespie, J. Kitzes, C. Marshall, N. Matzke, D.P. Mindell, E. Revilla, and A.B. Smith. 2012. Approaching a state shift in Earth's biosphere. *Nature,* 486: 52-58.

Bassil, K.L., C. Vakil, M. Sanborn, D.C. Cole, J.S. Kaur and K.J. Kerr. 2007. Cancer health effects of pesticides: systematic review. *Canadian Family Physician,* 53(10): 1704-1711.

Bateson, G. 1972. *Steps to an ecology of mind.* New York: Chandler.

Belachew, O., and M.I. Zuberi. 2015. Perception of climate change and livelihood of a farming community of Maruf Kebele, Central Oromia, Ethiopia. *American Journal of Climate Change,* 4: 269-281. DOI: 10.4236/ajcc.2015.43022

Bell, J., M. Brubaker, K. Graves, and J. Berner. 2010. Climate change and mental health: uncertainty and vulnerability for Alaska natives. *Center for Climate and Health Bulletin,* 3. Retrieved from http://anthc.org/wp-content/uploads/2016/01/CCH-Bulletin-No-3-Mental-Health.pdf

Bhutiyani, M.R., V.S. Kale, and N.J. Pawar. 2007. Long term trends in maximum, minimum and mean annual temperatures across the north-western Himalaya during the twentieth century. *Climate Change,* 85: 159-177.

Brown, D., R.R. Chanakira, K. Chatiza, M. Dhliwayo, D. Dodman, M. Masiiwa, D. Muchdenyika, P. Mugabe, and S. Zvigadza. 2012. Climate change impacts, vulnerability and adaptationin Zimbabwe. *IIED Climate Change Working Paper No. 3.* Retrieved from https://pubs.iied.org/sites/default/files/pdfs/migrate/10034IIED.pdf

Buechler, S. 2009. Gender, water, and climate change in Sonora, Mexico: implications for policies and programmes on agricultural income-generation. *Gender & Development,* 17(1): 51-66.

Burns, R.B. 2000. *Introduction to research methods* (4th ed). London: Sage Publications.

Campbell, B.M., S.J. Vermeulen, P.K. Aggarwal, C. Corner-Dolloff, E. Girvetz, A. M. Loboguerrero, J. Ramirez-Villegas, T. Rosenstock, L. Sebastian, P. Thornton, and E. Wollenberg. 2016. Reducing risks to food security from climate change. *Global Food Security,* 11(2016): 34-43.https://doi.org/10.1016/j.gfs.2016.06.002

Chambers, R. and G. Conway. 1991. Sustainable rural livelihoods: practical concepts for the 21st century. *IDS Discussion Paper,* 296. Retrieved from https://www.ids.ac.uk/download.php?file=files/Dp296.pdf

Cianconi P., S. Betrḍ, and L. Janiri. 2020. The impact of climate change on mental health: a systematic descriptive review. *Frontiers in Psychiatry,* 11: 74. doi: 10.3389/fpsyt.2020.00074

Connor, L. 2016. *Climate change and anthropos: planet, people and places*. London: Routledge.

Costello, A., M. Abbas, A. Allen, S. Ball, S. Bell, R. Bellamy, S. Friel, N. Groce, A. Johnson, M. Kett, M. Lee, C. Levy, M. Maslin, D. McCoy, B. McGuire, H. Montgomery, D. Napier, C. Pagel, J. Patel, J.A. Puppim de Oliveira, N. Redclift, H. Rees, D. Rogger, J. Scott, J. Stephenson, J. Twigg, J. Wolff, and C. Patterson. 2009. Managing the health effects of climate change: Lancet and University College London Institute for Global Health Commission. *The Lancet,* 373: 1693-1733. Retrieved from https://doi.org/10.1016/S0140-6736(09)60935-1

Crate, S.A., and M. Nuttall, eds. 2009. *Anthropology and climate change: from encounters to actions.* Walnut Creek, California: Left Coast Press.

———. 2016. *Anthropology and climate change: from encounters to actions* (2nd ed.). London: Routledge.

Darlington, Y., and D. Scott. 2002. *Qualitative research in practice: stories from the field.* Buckingham: Open University Press.

Dash, S.K., R.K. Jenamani, S.R. Kalsi, and S.K. Panda. 2007. Some evidence of climate change in twentieth-century India. *Climatic Change,* 85: 299-321. DOI 10.1007/s10584-007-9305-9

Datta, S. 2013. Impact of climate change in Indian horticulture-a review. *International Journal of Science, Environment and Technology,* 2 (4): 661-671. ISSN 2278-3687(O).

De, U.S., R.K. Dube, and G.S. Prakasa Rao. 2005. Extreme weather events over India in the last 100 years. *Journal of Indian Geophysical Union,* 9(3): 173-187.

Dhara, V.R., P.J. Schramm, and G. Luber. 2013. Climate change & infectious diseases in India: implications for health care providers. *The Indian Journal of Medical Research,* 138(6): 847-852.

District Disaster Management Authority Shimla. 2012. *District disaster management plan of District Shimla.* Himachal Pradesh: Himachal Pradesh State Disaster Management Authority.

Douglas, I., K. Alam, M.A. Maghenda, Y. Mcdonnell, L. McLean, and J. Campbell. 2008. Unjust waters: climate change, flooding and the urban poor in Africa. *Environment and Urbanization,* 20(1): 187-205.

Dove, M.R. 2014. Introduction: the anthropology of climate change six millennia of study of the relationship between climate and society. In *The anthropology of climate change: an historical reader*. M.R. Dove, ed. pp 41-46. UK: John Wiley and Sons, Inc.

EEA. 2015. *Soil and climate change.* Retrieved from https://www.eea.europa.eu/ signals/signals-2015/articles/soil-and-climate-change.

Ellis, F. 2000. The determinants of rural livelihood diversification in developing countries. *Journal of Agricultural Economics,* 51(2): 289-302. http://doi.org/10.1111/j.1477-9552.2000.tb01229.x.

Eriksen, T.H. 2021. Climate change. In *The Cambridge Encyclopaedia of Anthropology.* Retrieved from *http://doi.org/10.29164/21climatechange*

FAO. 2004. *Report twenty-seventh FAO regional conference for Asia and the Pacific.* Retrieved from http://www.fao.org/3/j2409e/j2409e00.htm

————. 2016. *Livestock and climate change.* Retrieved from http://www.fao.org/3/i6345e/i6345e.pdf

Fink, A. 2005. *Conducting research literature reviews: from the internet to paper.* Thousand Oaks,California: Sage Publications.

Gentle, P., and T.N. Maraseni. 2012. Climate change, poverty and livelihoods: adaptation practices by rural mountain communities in Nepal. *Environmental Science & Policy,* 21: 24-34.

Government of Himachal Pradesh. 2021. *Economic Survey 2020-21 Saving Lives and Livelihood.* Himachal Pradesh: Economic and Statistics Department.

Government of India. 2011. *Census of India 2011: Himachal Pradesh series-03 part XII-B district census handbook Shimla village and town-wise primary census abstract* (*PCA*). Himachal Pradesh: Directorate of Census Operations.

————. 2020. India 2020: a reference annual. New Delhi: Publication Division, Ministry of Information and Broadcasting, Government of India.

Goyal, M.K., and Rao Y. Surampalli. 2018. Impact of climate change on water resources in India. *Journal of Environmental Engineering,* 144(7): 04018054. DOI: 10.1061/(ASCE) EE.1943-7870.0001394

Gray, D.E. 2004. *Doing research in the real world.* London: Sage.

Green, D., L. Alexander, K. Mclnnes, J. Church, N. Nicholls, and N. White. 2010. An assessment of climate change impacts and adaptation for the Torres Strait Islands, Australia. *Climatic Change,* 102(3): 405-433.

Hassan, R., and C. Nhemachena. 2008. Determinants of African farmers' strategies for adapting to climate change: multinomial choice analysis. *African Journal of Agricultural and Resource Economics,* 2(1): 83-104.

Hayelom, B., Y. Chen, Z. Marsie, and M. Negash. 2017. Temperature and precipitation trend analysis over the last 30 years in southern Tigray regional state, Ethiopia. *Preprints* 2017020014. (doi: 10.20944/preprints201702.0014.v1) Retrieved from https://www.preprints.org/manuscript/201702.0014/v1

Hazarika, T.K. 2013. Climate change and Indian horticulture: opportunities, challenges and mitigation strategies. *International Journal of Environmental Engineering and Management,* 4(6): 629-630. ISSN 2231-1319.

Hendry, J. 2014. *Science and sustainability: learning from indigenous wisdom.* London: Palgrave Macmillan.

Hewitt, K., and M. Mehta. 2012. Rethinking risk and disasters in mountain areas. *Journal of Alpine Research,* 100(1). doi:10.4000/rga.1653.

Hingane, L.S., K. Rupa Kumar and B.V. Ramana Murthy. 1985. Long-term trends of surface air temperature in India. *Journal of Climatology,* 5: 521-528.

HP-SAPCC. 2012. *Himachal Pradesh state strategy and action plan on climate change.* Himachal Pradesh: Department of Environment, Science and Technology, Govt. of Himachal Pradesh

Huber-Sannwald, E., F.T. Maestre, J.E. Herrick, and J.F. Reynolds. 2006. Ecohydrological feedbacks and linkages associated with land degradation: a case study from Mexico. *Hydrological Processes,* 20: 3395-3411.

IFAD. 2009. *High food prices: impact and recommendations.* Retrieved from http://www.ifad.org/operations/food/ceb.htm

ILO. 2008. 97th session of the International Labour Conference. *Report IV. Promotion of rural employment for poverty reduction.* Geneva, Switzerland: ILO.

IMD. 2010. *Climate of Himachal Pradesh climatological summaries of state series no. 15.* New Delhi: Controller of Publications, Government of India, Department of Publication.

IPCC. 2007a. *Climate change 2007: impacts, adaptation and vulnerability. Contribution to the working group II to the fourth assessment report of Intergovernmental Panel on Climate Change,* M.L. Parry, O.F. Canziani, J.P. Palutik, P.J. van der Linden, and C.E. Hanson, eds. Cambridge: Cambridge University Press.

———. 2007b. *Climate change 2007: synthesis report. Contribution of Working Groups I, II and III to the fourth assessment report of the Intergovernmental Panel on Climate Change.* Core writing team, R.K. Pachauri, and A. Reisinger, eds. Geneva, Switzerland: IPCC.

———. 2012. Summary for policymakers. In *Managing the risks of extreme events and disasters to advance climate change adaptation.* C.B. Field, V. Barros, T.F. Stocker, D. Qin, D.J. Dokken, K.L. Ebi, M.D. Mastandrea, K.J. Mach, G.K. Plattner, S.K. Allen, M. Tignor, and P.M. Midgely, eds.

pp. 1-19. Cambridge, UK: Cambridge University Press.

————. 2014. *Climate change 2014: synthesis report. Contribution of Working Groups I, II and III to the fifth assessment report of the Intergovernmental Panel on Climate Change.* Core Writing Team, R.K. Pachauri, and L.A. Meyer, eds. Geneva, Switzerland: IPCC.

————. 2021. *Climate change 2021: the physical science basis. Contribution of working group I to the sixth assessment report of the Intergovernmental Panel on Climate Change.* V. Masson-Delmotte, P. Zhai, A. Pirani, S.L. Connors, C. Pīan, S. Berger, N. Caud, Y. Chen, L. Goldfarb, M.I. Gomis, M. Huang, K. Leitzell, E. Lonnoy, J.B.R. Matthews, T.K. Maycock, T. Waterfield, O. Yelekśi, R. Yu and B. Zhou, eds. Cambridge: Cambridge University Press.

IRP and UNDP-India. 2012. *Guidance note on recovery livelihood.* Japan: International Recovery Platform Secretariat.

Islam, T., and J. Ryan. 2016. Mitigation in the private sector. In *Hazard mitigation in emergency management.* T. Islam, J. Ryan, eds. pp. 101-124. UK: Butterworth-Heinemann. ISBN 978-0-12-420134-7. Retrieved from https://doi. org/10.1016/B978-0-12-420134-7.00004-7.

Iwasaki, S., B.H.N. Razafindrabe and R. Shaw. 2009. Fishery livelihoods and adaptation to climate change: a case study of Chilika lagoon, India. *Mitigation and Adaptation Strategies for Global Change,* 14(4): 339-355.

Jabeen, H., C. Johnson, and A. Allen. 2010. Built-in resilience: learning from grassroots coping strategies for climate variability. *Environment and Urbanization,* 22(2): 415-431.

Jain, S.K., and V. Kumar. 2012. Trend analysis of rainfall and temperature data for India. *Current Science,* 102(1): 37-49.

Jaswal, A.K., S.C. Bhan, A.S. Karandikar, and M.K. Gujar. 2015. Seasonal and annual rainfall trends in Himachal Pradesh during 1951-2005. *Mausam,* 66 (2): 247-264.

Kinyangi, J., Herrero, A. Omolo, J. Van de Steeg, and P. Thoronto. 2009. *Scoping study on vulnerability to climate change and climate variability in the Greater Horn of Africa: mapping impact and adaptive capacity.* Nairobi: ILRI.

Kothari, C. R. and G. Garg. 2014. *Research methodology methods and techniques* (3rd ed.). New Delhi: New Age International (P) Ltd.

Kulkarmi, J., and N. Leary. 2007. Climate change and vulnerability in developing country regions. In *Draft final report of the AIACC project, a global environmental facility enabling activity in the climate change focal area, project no GFL, 2328-2724-4330*. Nairobi, Kenya: UNEP. Retrieved from http://www.aiaccproject.org

Kvale, S. 1996. *Interviews: an introduction to qualitative research interviewing*. London: Sage.

Ladaniya, M.S. 2011. Climate change and its impact on fruit quality and post-harvest management strategy. In *Impact of climate change on fruit crops*. W. S. Dhillon and P. S. Aulakh, eds. pp. 385-402. Delhi: Narendra Publishing House.

Lake, I.R., L. Hooper, A. Abdelhamid, G. Bentham, A.B.A. Boxall, A. Draper, S. Fairweather-Tait, M. Hulme, P.R. Hunter, G. Nichols, and K.W. Waldron. 2012. Climate change and food security: health impacts in developed countries. *Environmental Health Perspectives,* 120: 1520-1526.http://dx.doi.org/10.1289/ehp.1104424

Lavrakas, P.J. 2008. *Encyclopedia of survey research methods*. Thousand Oaks, California: Sage Publications, Inc.doi: 10.4135/9781412963947).

Le, S.T., and C.D. Vo. 2021. The livelihood adaptability of households under the impact of climate change in the Mekong delta. *Journal of Agribusiness in Developing and Emerging Economies*, 11(1): 7-26. DOI 10.1108/JADEE-09-2019-0139

McCarthy, J.J., O.F. Canzlani, N.A. Leary, D.J. Dokken, and K.S. White, eds. 2001. *Climate change. 2001: impacts, adaptation, and vulnerability. Contribution of working group II to the third assessment report of the Intergovernmental Panel on Climate Change*. Cambridge: Cambridge University Press.

McMichael, A.J., and E. Lindgren. 2011. Climate change: present and future risks to health, and necessary responses (review). *Journal of Internal Medicine,* 270: 401-413. doi: 10.1111/j.1365-2796.2011.02415.x

Miles, M., and A.M. Huberman. 1994. *Qualitative data analysis: an expanded source book* (2nd ed.). Thousand Oaks, California: Sage.

Nelson, D.R., W. Neil Adger, and K. Brown. 2007. Adaptation to environmental change: contributions of a resilience framework. *Annual Review of Environment and Resources,* 32(1): 395-419. Retrieved from https://doi.org/10.1146/annurev.energy.32.051807.090348

Nightingale, A.J. 2020. Triangulation. In *International Encyclopedia of Human Geography* (2nd ed). A. Kobayashi, ed. pp. 477-480. ISBN 9780081022962. https://doi.org/10.1016/B978-0-08-102295-5.10437-8.

Transportation Research Board and National Research Council. 2008. *Potential impacts of climate change on U.S. transportation: special report 290.* Washington, DC: The National Academies Press.https://doi.org/10.17226/12179

Oduniyi, O.S. 2014. Climate change awareness: a case study of small-scale maize farmers in Mpumalanga province, South Africa. A dissertation submitted at the Department of Agriculture and Animal Health, University of South Africa.

Olsson, L., M. Opondo, P. Tschakert, A. Agrawal, S.H. Eriksen, S. Ma, L.N. Perch, and S.A. Zakieldeen. 2014. Livelihoods and poverty. In *Climate Change 2014: Impacts, adaptation, and vulnerability. Part A: Global and Sectoral Aspects. Contribution of Working Group II to the Fifth Assessment Report of the Intergovernmental Panel on Climate Change.* C.B. Field, V.R. Barros, D.J. Dokken, K.J. Mach, M.D. Mastrandrea, T.E. Bilir, M. Chatterjee, K.L. Ebi, Y.O. Estrada, R.C. Genova, B. Girma, E.S. Kissel, A.N. Levy, S. MacCracken, P.R. Mastrandrea, and L.L.White. eds. pp. 793-832. Cambridge, United Kingdom: Cambridge University Press.

Orlove, B., J.C.H. Chiang, and M.A. Cane. 2000. Forecasting Andean rainfall and crop yield from the influence of El Niño on Pleiades visibility. *Nature,* 403: 68-71.

Pandve, H.T. 2009. India's national action plan on climate change. *Indian Journal of Occupational and Environmental Medicine,* 13(1): 17. Retrieved from https://doi.org/10.4103/0019-5278.50718.

Pareek, N. 2017. Climate change impact on soils: adaptation and mitigation. *MOJ Ecology & Environmental Science,* 2(3): 136-139. DOI: 10.15406/mojes.2017.02.00026

Randell, H., and C. Gray. 2019. Climate change and educational attainment in the global tropics. *Proceedings of the National Academy of Sciences.* DOI: 10.1073/pnas.1817480116

Randev, A.K. 2009. Impact of climate change on apple productivity in Himachal Pradesh, India. Paper presented at *60th International Executive Council Meeting & 5th Asian Regional Conference,* 6-11 December 2009. New Delhi, India.

Rathore, L.S., S.D. Attri, and A.K. Jaswal. 2013. State level climate change trends in India. *Meteorological Monograph* No. ESSO/IMD/EMRC/02/2013

Rautela, P., and B. Karki. 2015. Impact of climate change on life and livelihood of indigenous people of higher Himalaya in Uttarakhand, India. *American Journal of Environmental Protection,* 3(4): 112-124. doi: 10.12691/env-3-4-2.

Rayner, S., and E. Malone, eds. 1998. *Human Choice and Climate Change.* Columbus: Battelle Press.

Rescia, A., A. Pons, I. Lomba, C. Esteban, and J. Dover. 2008. Reformulating the social-ecological system in a cultural rural mountain landscape in the Picos de Europa region (northern Spain). *Landscape and Urban Planning,* 88 (1): 23-33.

Ruby, J. 1996. Visual anthropology. In *Encyclopedia of Cultural Anthropology* (Vol. 4). D. Levinson, and M. Ember, eds. pp. 1345-1351. New York: Henry Holt.

Saadat, M.A., and A.K.M. Saiful Islam. 2011. Impact of climate change on rural livelihood: a case study. Paper presented at *3rd International Conference on Water & Flood Management,* 8-10 January, 2011. Dhaka, Bangladesh.

Salve, Prachi. 2013. How India's construction workers get gypped of their due. *India Spend.* Retrieved from http://www.indiaspend.com/investigations/how-indias-constructionworkers-get-gypped-of-their-due-51414

Sarma, K.V.S. 2001. *Statistics made simple do it yourself on PC.* New Delhi: Prentice-Hall of India Private Limited.

Save the Children. 2008. Education and climate change in the Middle East and North Africa. *A report from the World*

Bank on working for a world free of poverty. Retrieved from http://blogs.worldbank.org/arabvoices/education-and-climate-change-middle-east-andnorthafrica

Schneider, S.H., S. Semenov, A. Patwardhan, I. Burton, C.H.D. Magadza, M. Oppenheimer, A.B. Pittock, A. Rahman, J.B. Smith, A. Suarez, and F. Yamin. 2007. Assessing key vulnerabilities and the risk from climate change. In *Climate Change 2007: Impacts, Adaptation and Vulnerability. Contribution of Working Group II to the Fourth Assessment Report of the Inter-governmental Panel on Climate Change.* M.L. Parry, O.F. Canziani, J.P. Palutikof, P.J. van der Linden, and C.E. Hanson, eds. pp. 779-810. Cambridge, UK: Cambridge University Press.

Scoones, I. 2009. Livelihoods perspectives and rural development. *The Journal of Peasant Studies,* 36(1): 171-196. DOI: 10.1080/03066150902820503

Sharma, D.K. 2011. A brief outline and macro-dimensions of population in Himachal Pradesh vis-à-vis other states in the region. In *Hill post*. Retrieved from https://hillpost.in/2011/04/a-brief-outline-and-macro-dimensions-of-population-in-himachal-pradesh-vis-à-vis-other-states-in-the-region/27184/ accessed on 16th August, 2021.

Singh, R., C. Pal. 1961. *Census 1961 Himachal Pradesh District Handbook 4 Mahasu*. Himachal Pradesh: Superintendent of Census Operations.

Stafford, J., G. Wendler, and J. Curtis. 2000. Temperature and precipitation of Alaska: 50 year trend analysis. *Theoretical and Applied Climatology,* 67: 33-44. https://doi.org/10.1007/s007040070014

Steffen, W., A. Persson, L. Deutsch, J. Zalasiewicz, M. Williams, K. Richardson, C. Crumley, P. Crutzen, C. Folke, L. Gordon, M. Molina, V. Ramanathan, J. Rockstrṭm, M. Scheffer, H.J. Schellnhuber, and U. Svedin. 2011. The anthropocene: from global change to planetary stewardship. *Ambio,* 40(7): 739-761. https://doi.org/10.1007/s13280-011-0185-x

Steward, J. 1955. *Theory of Culture Change.* Urbana: University of Illinois Press.

Straza T., S. Lui, and B. Burfitt. 2018. Effects of climate change on society, culture and gender relevant to the Pacific islands.

Pacific Marine Climate Change Report Card Science Review 2018: 201-210.

Stringer, L.C., C. Twyman, and D.S.G. Thomas, 2007. Learning to reduce degradation on Swaziland's arable land: enhancing understandings of Striga Asiatica. *Land Degradation and Development,* 18(2): 163-177.

TERI. 2015. *Green Growth and Climate Change Mitigation in Himachal Pradesh*. New Delhi: The Energy and Resources Institute.

Tacchi, J., D. Slater, and G. Hearn. 2003. *Ethnographic Action Research: a User's Handbook.* New Delhi: UNESCO

Tandzi, L.N., and C.S. Mutengwa. 2020. Factors affecting yield of crops. In *Agronomy-climate Change and Food Security*. Dr. Amanullah, and S. Khalid eds. London, UK: Intechopen. DOI: 105772/intechopen.90672.

Thomas, D.S.G., C. Twyman, H. Osbahr, and B. Hewitson. 2007. Adaptation to climate change and variability: farmer responses to intra-seasonal precipitation trends in South Africa. *Climatic Change,* 83(3): 301-322.

The New Encyclopaedia Britannica. 2002. Horticulture. In *The New Encyclopaedia Britannica,* Volume 6. USA: Encyclopaedia Britannica, Inc.

The Oxford English Dictionary. 1989. Horticulture. In *The Oxford English Dictionary* (2nd ed), Volume VII. J.A. Simpson and E.S.C. Weiner, eds. p. 412. Oxford: Clarendon Press.

Tompkins, L., W. Adger, E. Boyd, S. Nicholson-Cole, K. Weatherhead, and N. Arnell. 2010. Observed adaptation to climate change: UK evidence of transition to a well-adapting society. *Global Environmental Change,* 20 (4): 627-635.

Toulmin, C., R. Leonard, K. Brock, N. Coulibaly, G.Carswell, and D. Dea. 2000. Diversification of livelihoods: evidence from Mali and Ethiopia. *IDS Research Report,* 47:59. Brighton, UK: Institute Development Studies, University of Sussex.

Tuncok, K. 2021. Impacts of climate change in Central Asia. *Academia Letters, Article 1219*. https://doi.org/10.20935/AL1219.

Tylor, E.B. 1871. *Primitive Culture: Researches into the Development of Mythology, Philosophy, Religion, Art and Custom*. London: John Murray, Albermarle Street.

UNFCCC. 2011. *Fact sheet: climate change science—the status of climate change science today*. Retrieved from https://unfccc.int/files/press/backgrounders /application/pdf/press_factsh_science.pdf

UNICEF. 2008. The impact of climate change on education. *A report from the World Bank on working for a world free of poverty*. Retrieved from http://blogs.worldbank.org/arabvoices/education-and-climate-changemiddleeast -and-north-africa.

UNISDR/SDMC. 2014. *Integration of disaster risk reduction and climate change adaptation in SAARC region: implementation of the Thimphu statement on climate change—a comprehensive study of the policy, institutional landscape, and resource allocation for disaster risk reduction and climate change adaptation in South Asia* (*Disaster prevention, preparedness & management, linkages with CCA*). New Delhi: SDMC.

USGCRP. 2014. *Climate change impacts in the United States: the third national climate assessment.* J.M. Melillo, T.C. Richmond, and G.W. Yohe, eds. Washington, DC: U.S. Global Change Research Program.

————. 2016. *The impacts of climate change on human health in the United States: a scientific assessment*. A. Crimmins, J. Balbus, J.L. Gamble, C.B. Beard, J.E. Bell, D. Dodgen, R.J. Eisen, N. Fann, M.D. Hawkins, S.C. Herring, L. Jantarasami, D.M. Mills, S. Saha, M.C. Sarofim, J. Trtanj, and L. Ziska, eds. Washington, DC: U.S. Global Change Research Program. http://dx.doi.org/10.7930/J0R49NQX

Vidaurre de la Riva, M., A. Lindner, and J. Pretzsch. 2013. Assessing adaptation—climate change and indigenous livelihood in the Andes of Bolivia. *Journal of Agriculture and Rural Development in the Tropics and Subtropics,* 114(2):109-122. Retrieved from https://www.jarts.info/index.php/jarts/article/view/201308 1343342

Weart, S.R. 2004. *The Discovery of Global Warming*. Cambridge: Harvard University Press. Retrieved from http://www.aip.org/history/climate/index.html

White, L. 1949. *The Science of Culture: A Study of Man and Civilization.* New York: Grove Press.

Williams, D.K. 2021. Differences between north-facing and south-facing slopes. *Sciencing.com*. Retrieved from https://sciencing.com/differences-between-north-southfacing-slopes-8568075.html.

Willis, K. 2006. Interviewing. In *Doing Development Research*. V. Desai, and R. Potter, eds. pp. 144-152. London: Sage.

WMO. 2021. *State of the Global Climate 2020.WMO-No. 1264*. Geneva, Switzerland: WMO.

Xu, J., R.E. Grumbine, A. Shrestha, M. Eriksson, X. Yang, Y. Wang, and A. Wilkes. 2009. The Melting Himalayas: Cascading effects of climate change on water, biodiversity, and livelihoods. *The Journal of the Society for Conservation Biology,* 23(3): 520-530. https://doi.org/10.1111/j.1523-1739.2009.01237.x

Yin, R.K. 2003. *Case Study Research: Design and Methods*. Thousand Oaks, California: Sage Publications, Inc.

Internet Sources

http://himachalpr.gov.in/OneNews.aspx?Language=1&ID=18224

http://himachalpr.gov.in/PressReleaseByYear.aspx?Language=1&ID=21757&Type=2&Date=02/05/2021

http://www.hpccc.gov.in/Default.aspx

http://www.shimlamc.org/MC/admin/Pages/page/CDPChap5.aspx

https://censusindia.gov.in/census_and_you/economic_activity.aspx

https://jicahpforestryproject.com/about/rationale

https://www.greengrowthknowledge.org/project/himachal-pradesh-mid-himalayan-watershed-development-project

https://www.sciencedirect.com/science/article/pii/B9780081022955104378

Index

Adaptation strategies and Government's efforts to combat climate change, 175-198
  adaptation strategies acquired by the people for
    the high-altitude village Jashla, 186
    the lower altitude village Bhawana, 176
    the lower-middle altitude village Kiari, 180
    the upper-middle altitude village Kufar Bag, 182
  adaptation strategies acquired by the people to overcome the impact of climate change, 175
  government efforts being taken to combat and mitigate climate change, 192
    issues related to the schemes faced by the respondents, 195
    schemes and projects being implemented in the field area as per the respondents, 194
Adaptation strategies, 191
Adaptation, 176
Agricultural livelihood, impact of climate change on, 24
Anthropology and climate change, 14
Anthropology of climate change, 3
  history of anthropology of climate change, 4
  importance of anthropological approach in studying climate change, 5
Anti-hail guns, 183
Anti-hail nets, 187
Asian Development Bank, 20

Bhawana village, 60
Biodiversity,
  changes in the in the field area (Fauna), 138
  changes in the in the field area (Flora), 137

Case study, 34, 35
Cash crop economy, 210
Cause-and-effect relationship, 218
Central Asia, 125
Climate change and livelihood,
  nexus between, 8
  of the area of study, 9
Climate change and people's perception, 65
  analysis of climatic data of the study area, 65
    rainfall, 68

snowfall, 74
temperature, 65
comparison of people's perception and climate data, 91
factors affecting the perception about climate change, 85
impact of climate change on day-to-day lives, 88
people's perception about cause of climate change, 81
people's perception about climate change and its impact on their day-to-day lives, 76
rainfall, 77
snowfall, 79
temperature, 76
Climate change, 1, 209
adaptation made to combat and mitigate climate change, 25
changing dynamics of social relations as an adaptation to combat, 188
climate change trends, 15
impact of climate change on agricultural livelihood, 4
impact of climate change on livelihood, 18
steps to mitigate the impact of, 190
Cool temperature, 163, 211
Crop calendar, 202
Crop varieties, calendar and crop production, impact of climate change on, 93-124
crop calendar, 97
crop documentation, 94
changes in crop varieties with changing climate, 95
crop production, 105
crop production based on secondary data, 109
people's perception, 105
Culture, 175

Ethnoclimatological research, 15
European Environment Agency (EEA), 126

Factors affecting the perception about climate change, 85
education, 86
modernisation and globalisation, 85
traditional and scientific theories, 86
Financial capital, impact of climate change on, 22
Focus Group Discussions (FGDs), 35, 200
Food and Agriculture Organization (FAO), 21
Fruit crop, productivity of, 160
Fruit industry, 211

Global warming, 132
Grassland fires, 141

High-Density Plant (HDP), 151, 177, 182, 184, 191, 211, 212, 213
Himachal Pradesh Mid-Himalayan Watershed Development Project, 193
Himachal Pradesh, 9, 47
administrative units, 49
climate, 51
demography, 50
economy, 50
topography, 48
Himalayan regions, 162
Horticulture, 93, 165
evolution of the present horticulture practices, 95
Horticulture-based economy, 13

Human capital, impact of climate change on, 19
education, 19
employment, 20
health, 20
household members, 19

India, 9, 46
Indian Meteorological Department (IMD), 47
Intergovernmental Panel on Climate Change (IPCC), 1, 8, 11, 16, 199
International Fund for Agricultural Development (IFAD), 22

Japan International Cooperation Agency (JICA), 192, 194, 215
Jashla, 63, 206

Kiari village, 61
Kotkhai tehsil, 9, 10, 54
apple production, production area and productivity, 109
apricot production, production area and productivity, 109
cherry production, production area and productivity, 121
demography, 55
dress and ornament, 59
economy, 56
family structure, 59
field area, 59
Bhawana, 60
Jashla, 63
Kiari, 61
Kufar Bag, 62
history, 55
language, 59
peach production, production area and productivity, 116
pear production, production area and productivity in, 112
religion, 57
rituals, customs and festivals, 58
Krishi Utpadan Sarankshan Yojana, 197
Kufar Bag, 62

Lift Water Supply Scheme (LWSS), 194
Livelihood assets, 217
Livelihood contexts, 7
Livelihood, 5
Livelihood--impact of climate change on, 125-174
financial assets of livelihood, 158
horticulture, 159
extreme weather events, 169
increase in use of chemicals, 172
rainfall, 164
snowfall, 167
temperature, 162
human assets of livelihood, 141
active labour, 143
education, 145
health, 145
household members, 142
knowledge and skills, 151
natural assets of livelihood, 127
biodiversity, 136
soil/land, 127
water, 129
physical assets of livelihood, 154
social assets of livelihood, 152

Marriage, 209
Microsoft Excel (MS Excel), 38

National Action Plan on Climate Change (NAPCC), 192

Natural assets of livelihood, 160
Natural capital, impact of climate change on, 21
  soil, 22
  water, 21

Parvat Dhara scheme, 193
Photography, 36
Physical capital, impact of climate change on, 21
Plantation drives,193
Present study (the),
  aim and objectives, 27
  area and people, 46-64
  chapterization, 44
  data analysis, 37
  data collection, 33
  details of field visits, 32
  ethical considerations, 32
  introduction, 1-45
  limitations of the study, 44
  rationale of the problem, 12
  rationale of the selection of study area, 28
  research methodology, 27
  review of literature, 13
  sampling, 30
  significance of the study, 43
  statement of the problem, 11
  summary and conclusion, 199-218
    conclusion, 216
    summary, 199
  universe and unit of study, 28

Quasi participant observation, 36

Rainwater Harvesting Structures (RWHS), 193

SAARC region, 22
Shimla, 51
  administrative units, 52
  climate, 53
  demography, 52
  economy, 53
  topography, 52
Social capital, impact of climate change on, 23
Social relations, 188

The Energy and Resources Institute (TERI), 9
Types of livelihood assets, 6

U.S. Global Change Research Program, 20
United Nations Framework Convention on Climate Change (UNFCCC), 2

Water conservation, 181
Water, 205
Watershed Development Project, 196, 215
World Meteorological Organization (WMO), 2, 16, 20